COMPUTER NUMERICAL CONTROL PROGRAMMING OF MACHINES

Larry Horath

Merrill, an imprint of
Macmillan Publishing Company
New York

Maxwell Macmillan Canada
Toronto

Maxwell Macmillan International
New York   Oxford   Singapore   Sydney

Editor: Stephen Helba
Production Editor: Louise N. Sette
Art Coordinator: Peter A. Robison
Cover Designer: Cathleen Norz
Production Buyer: Pamela D. Bennett
Illustrations: Precision Graphics

This book was set in Times Roman by Compset, Inc. and was printed and bound by Book Press, Inc., a Quebecor America Book Group Company. The cover was printed by Phoenix Color Corp.

Copyright © 1993 by Macmillan Publishing Company, a division of Macmillan, Inc. Merrill is an imprint of Macmillan Publishing Company.

Printed in the United States of America

All rights reserved. No part of this book may be reproduced or transmitted in any form or by any means, electronic or mechanical, including photocopy, recording, or any information storage and retrieval system, without permission in writing from the Publisher.

The Publisher offers discounts on this book when ordered in bulk quantities. For more information, write to: Special Sales Department, Macmillan Publishing Company, 445 Hutchinson Avenue, Columbus, OH 43235, or call 1-800-228-7854

Macmillan Publishing Company
866 Third Avenue
New York, NY 10022

Macmillan Publishing Company is part of the
Maxwell Communication Group of Companies.

Maxwell Macmillan Canada, Inc.
1200 Eglinton Avenue East, Suite 200
Don Mills, Ontario M3C 3N1

Library of Congress Cataloging-in-Publication Data

Horath, Larry.
   Computer numerical control programming of machines / by Larry Horath.
      p.   cm.
   Includes index.
   ISBN 0-02-357201-9
    1. Machine—tools—Numerical control. I. Title.
TJ1189.H63   1993
621.9'023—dc20   92-29570
                                                                                                     CIP

Printing: 1 2 3 4 5 6 7 8 9   Year: 3 4 5 6 7

MERRILL'S INTERNATIONAL SERIES IN ENGINEERING TECHNOLOGY

INTRODUCTION TO ENGINEERING TECHNOLOGY

Pond, *Introduction to Engineering Technology*, 2nd Edition, 0-02-396031-0

ELECTRONICS TECHNOLOGY

Electronics Reference

Adamson, *The Electronics Dictionary for Technicians*, 0-02-300820-2
Berlin, *The Illustrated Electronics Dictionary*, 0-675-20451-8
Reis, *Becoming an Electronics Technician: Securing Your High-Tech Future*, 0-02-399231-X

DC/AC Circuits

Boylestad, *DC/AC: The Basics*, 0-675-20918-8
Boylestad, *Introductory Circuit Analysis*, 6th Edition, 0-675-21181-6
Ciccarelli, *Circuit Modeling: Exercises and Software*, 2nd Edition, 0-02-322455-X
Floyd, *Electric Circuits Fundamentals*, 2nd Edition, 0-675-21408-4
Floyd, *Electronics Fundamentals: Circuits, Devices, and Applications*, 2nd Edition, 0-675-21310-X
Floyd, *Principles of Electric Circuits*, 4th Edition, 0-02-338531-6
Floyd, *Principles of Electric Circuits: Electron Flow Version*, 3rd Edition, 0-02-338501-4
Keown, *PSpice and Circuit Analysis*, 0-675-22135-8
Monssen, *PSpice with Circuit Analysis*, 0-675-21376-2
Tocci, *Introduction to Electric Circuit Analysis*, 2nd Edition, 0-675-20002-4

Devices and Linear Circuits

Berlin & Getz, *Fundamentals of Operational Amplifiers and Linear Integrated Circuits*, 0-675-21002-X
Berube, *Electronic Devices and Circuits Using MICRO-CAP II*, 0-02-309160-6
Berube, *Electronic Devices and Circuits Using MICRO-CAP III*, 0-02-309151-7
Bogart, *Electronic Devices and Circuits*, 3rd Edition, 0-02-311701-X
Tocci, *Electronic Devices: Conventional Flow Version*, 3rd Edition, 0-675-21150-6
Floyd, *Electronic Devices*, 3rd Edition, 0-675-22170-6
Floyd, *Electronic Devices: Electron Flow Version*, 0-02-338540-5
Floyd, *Fundamentals of Linear Circuits*, 0-02-338481-6
Schwartz, *Survey of Electronics*, 3rd Edition, 0-675-20162-4
Stanley, *Operational Amplifiers with Linear Integrated Circuits*, 2nd Edition, 0-675-20660-X
Tocci & Oliver, *Fundamentals of Electronic Devices*, 4th Edition, 0-675-21259-6

Digital Electronics

Floyd, *Digital Fundamentals*, 4th Edition, 0-675-21217-0
McCalla, *Digital Logic and Computer Design*, 0-675-21170-0
Reis, *Digital Electronics through Project Analysis* 0-675-21141-7
Tocci, *Fundamentals of Pulse and Digital Circuits*, 3rd Edition, 0-675-20033-4

Microprocessor Technology

Antonakos, *The 68000 Microprocessor: Hardware and Software Principles and Applications*, 2nd Edition, 0-02-303603-6
Antonakos, *An Introduction to the Intel Family of Microprocessors: A Hands-On Approach Utilizing the 8088 Microprocessor*, 0-675-22173-0
Brey, *The Advanced Intel Microprocessors*, 0-02-314245-6
Brey, *The Intel Microprocessors: 8086/8088, 80186, 80286, 80386, and 80486: Architecture, Programming, and Interfacing*, 2nd Edition, 0-675-21309-6
Brey, *Microprocessors and Peripherals: Hardware, Software, Interfacing, and Applications*, 2nd Edition, 0-675-20884-X
Gaonkar, *Microprocessor Architecture, Programming, and Applications with the 8085/8080A*, 2nd Edition, 0-675-20675-6

Gaonkar, *The Z80 Microprocessor: Architecture, Interfacing, Programming, and Design*, 2nd Edition, 0-02-340484-1
Goody, *Programming and Interfacing the 8086/8088 Microprocessor: A Product-Development Laboratory Process*, 0-675-21312-6
MacKenzie, *The 8051 Microcontroller*, 0-02-373650-X
Miller, *The 68000 Family of Microprocessors: Architecture, Programming, and Applications*, 2nd Edition, 0-02-381560-4
Quinn, *The 6800 Microprocessor*, 0-675-20515-8
Subbarao, *16/32 Bit Microprocessors: 68000/68010/68020 Software, Hardware, and Design Applications*, 0-675-21119-0

Electronic Communications

Monaco, *Introduction to Microwave Technology*, 0-675-21030-5
Monaco, *Preparing for the FCC Radio-Telephone Operator's License Examination*, 0-675-21313-4
Schoenbeck, *Electronic Communications: Modulation and Transmission*, 2nd Edition, 0-675-21311-8
Young, *Electronic Communication Techniques*, 2nd Edition, 0-675-21045-3
Zanger & Zanger, *Fiber Optics: Communication and Other Applications*, 0-675-20944-7

Microcomputer Servicing

Adamson, *Microcomputer Repair*, 0-02-300825-3
Asser, Stigliano, & Bahrenburg, *Microcomputer Servicing: Practical Systems and Troubleshooting*, 2nd Edition, 0-02-304241-9
Asser, Stigliano, & Bahrenburg, *Microcomputer Theory and Servicing*, 2nd Edition, 0-02-304231-1

Programming

Adamson, *Applied Pascal for Technology*, 0-675-20771-1
Adamson, *Structured BASIC Applied to Technology*, 2nd Edition, 0-02-300827-X
Adamson, *Structured C for Technology*, 0-675-20993-5
Adamson, *Structured C for Technology (with disk)*, 0-675-21289-8
Nashelsky & Boylestad, *BASIC Applied to Circuit Analysis*, 0-675-20161-6

Instrumentation and Measurement

Berlin & Getz, *Principles of Electronic Instrumentation and Measurement*, 0-675-20449-6
Buchla & McLachlan, *Applied Electronic Instrumentation and Measurement*, 0-675-21162-X
Gillies, *Instrumentation and Measurements for Electronic Technicians*, 2nd Edition, 0-02-343051-6

Transform Analysis

Kulathinal, *Transform Analysis and Electronic Networks with Applications*, 0-675-20765-7

Biomedical Equipment Technology

Aston, *Principles of Biomedical Instrumentation and Measurement*, 0-675-20943-9

Mathematics

Monaco, *Essential Mathematics for Electronics Technicians*, 0-675-21172-7
Davis, *Technical Mathematics*, 0-675-20338-4
Davis, *Technical Mathematics with Calculus*, 0-675-20965-X

INDUSTRIAL ELECTRONICS/INDUSTRIAL TECHNOLOGY

Bateson, *Introduction to Control System Technology*, 4th Edition, 0-02-306463-3
Fuller, *Robotics: Introduction, Programming, and Projects*, 0-675-21078-X
Goetsch, *Industrial Safety and Health: In the Age of High Technology*, 0-02-344207-7
Goetsch, *Industrial Supervision: In the Age of High Technology*, 0-675-22137-4
Horath, *Computer Numerical Control Programming of Machines*, 0-02-357201-9
Hubert, *Electric Machines: Theory, Operation, Applications, Adjustment, and Control*, 0-675-20765-7
Humphries, *Motors and Controls*, 0-675-20235-3
Hutchins, *Introduction to Quality: Management, Assurance, and Control*, 0-675-20896-3
Laviana, *Basic Computer Numerical Control Programming*, 0-675-21298-7
Reis, *Electronic Project Design and Fabrication*, 2nd Edition, 0-02-399230-1
Rosenblatt & Friedman, *Direct and Alternating Current Machinery*, 2nd Edition, 0-675-20160-8
Smith, *Statistical Process Control and Quality Improvement*, 0-675-21160-3
Webb, *Programmable Logic Controllers: Principles and Applications*, 2nd Edition, 0-02-424970-X
Webb & Greshock, *Industrial Control Electronics*, 2nd Edition, 0-02-424864-9

MECHANICAL/CIVIL TECHNOLOGY

Keyser, *Materials Science in Engineering*, 4th Edition, 0-675-20401-1
Kraut, *Fluid Mechanics for Technicians*, 0-675-21330-4
Mott, *Applied Fluid Mechanics*, 3rd Edition, 0-675-21026-7
Mott, *Machine Elements in Mechanical Design*, 2nd Edition, 0-675-22289-3
Rolle, *Thermodynamics and Heat Power*, 3rd Edition, 0-675-21016-X
Spiegel & Limbrunner, *Applied Statics and Strength of Materials*, 0-675-21123-9
Wolansky & Akers, *Modern Hydraulics: The Basics at Work*, 0-675-20987-0
Wolf, *Statics and Strength of Materials: A Parallel Approach to Understanding Structures*, 0-675-20622-7

DRAFTING TECHNOLOGY

Cooper, *Introduction to VersaCAD*, 0-675-21164-6
Goetsch & Rickman, *Computer-Aided Drafting with AutoCAD*, 0-675-20915-3
Kirkpatrick & Kirkpatrick, *AutoCAD for Interior Design and Space Planning*, 0-02-364455-9
Kirkpatrick, *The AutoCAD Book: Drawing, Modeling, and Applications*, 2nd Edition, 0-675-22288-5
Kirkpatrick, *The AutoCAD Book: Drawing, Modeling, and Applications Including Version 12*, 3rd Edition, 0-02-364440-0
Lamit & Lloyd, *Drafting for Electronics*, 2nd Edition, 0-02-367342-7
Lamit & Paige, *Computer-Aided Design and Drafting*, 0-675-20475-5
Maruggi, *Technical Graphics: Electronics Worktext*, 2nd Edition, 0-675-21378-9
Maruggi, *The Technology of Drafting*, 0-675-20762-2
Sell, *Basic Technical Drawing*, 0-675-21001-1

TECHNICAL WRITING

Croft, *Getting a Job: Resume Writing, Job Application Letters, and Interview Strategies*, 0-675-20917-X
Panares, *A Handbook of English for Technical Students*, 0-675-20650-2
Pfeiffer, *Proposal Writing: The Art of Friendly Persuasion*, 0-675-20988-9
Pfeiffer, *Technical Writing: A Practical Approach*, 0-675-21221-9
Roze, *Technical Communications: The Practical Craft*, 0-675-20641-3
Weisman, *Basic Technical Writing*, 6th Edition, 0-675-21256-1

Dedication

With my sincerest love and devotion, I dedicate this text to my mother and father, Duane and Sandra, who taught me the value of hard work, patience, and dedication.

Preface

This text is designed to provide the student with a basic understanding of the concepts and procedures associated with numerical control (NC) and computer numerical control (CNC) technologies used in today's manufacturing industries. Material is presented in an easy-to-read, easy-to-understand format that does not assume that the student has a background in trigonometry or CNC programming. It is assumed, however, that the student has some basic math and machining background. This text is aimed at the training of both two-year technicians and four-year technologists.

Numerous examples are provided to support the fundamental points made in the text. A straightforward four-step process is used throughout the text to integrate the necessary theoretical background into practical, hands-on experience. The outcome of this four-step process is the knowledge necessary to understand, conduct, design, and develop CNC part programs and implementations on the job. The four-step process contains the following components: (1) clear explanations accompanied by illustrations; (2) step-by-step procedures; (3) applications designed to reinforce concepts; and (4) summary problems and examples designed to tie concepts together.

It is unrealistic to expect one text to cover every possible machine/controller combination. Since programming formats and techniques vary a great deal between machines and manufacturers, the student should supplement and compare the information provided in this text with the manual accompanying the machine to be programmed. Prior to programming a machine, the novice programmer should become acquainted with the capabilities and limitations of the machines they will be working with. Safety and safe working procedures cannot be overemphasized.

As the popularity of CNC technology and the application of CNC technology increases, the need for better basic instruction in the principles of numerical control will also increase. Based on this recognized need, the primary objectives of this text are to explain the "why" and demonstrate the "how" of CNC processes and programming procedures.

A glossary defines many of the terms associated with CNC technology and application. Also included in the glossary are nontechnical terms the reader may or may not be familiar with, but which are required for an adequate understanding of the material. Formulas, derivations, and sample calculations are provided at the point of application in the chapters and also in the appendices.

This text is meant as an introduction to the technology and programming of CNC; it is not a substitute for actual programming experience. So keep in mind that the material presented in this text is intended to guide and prepare you for independent programming. Do not become frustrated if something does not work for you the first time. Patience and practice will often provide a clear understanding of the material.

Acknowledgments

I would like to acknowledge and thank the following individuals for their contributions and assistance in the development of this manuscript: F. Gary Amy, University of Southwestern Louisiana; Phillip Foster, University of North Texas; James R. Drake, Cuyahoga Community College; Morris Ellenberg, Indiana Vocational Technical College; Dennis Kerns, Indiana Vocational Technical College; Jerry McDonald, Columbus State Community College; Kenneth Rennels, Purdue University-Indianapolis; Richard Neuverth, Hawkeye Institute of Technology; Paul Demers, University of Cincinnati; and Mark Bannatyne, Utah State University.

Contents

**Chapter 1 Introduction to
Computer Numerical Control** 1
   1.1   Introduction   1
   1.2   Numerical Control and Computer Numerical Control   2
   1.3   Part Programming and Input Media   6
   1.4   History of Numerical Control   10
   1.5   Drives and Control Loops   13
   1.6   Advantages and Disadvantages of Numerical Control   14
   1.7   FMS, CAD/CAM, and CIM   15
   1.8   Summary   17

**Chapter 2 Tooling Features for
Milling and Turning Machines** 19
   2.1   Introduction   19
   2.2   Cutting Speeds, Spindle Speeds, and Feed Rates   20
   2.3   Cutting Tool Materials   25
   2.4   Carbide Insert Terminology and Applications   28
   2.5   Tooling Systems   37
   2.6   Tool Lengths and Tool Length Offsets   42
   2.7   Tool Radius or Tool Diameter Compensation   46
   2.8   Adaptive Control   48
   2.9   Summary   49

Chapter 3 Programming Considerations 55
   3.1   Introduction   55
   3.2   Cartesian Coordinate System   57
   3.3   Axes of Machine Movement   57
   3.4   Point-to-Point and Linear-Cut Programming   60
   3.5   Continuous Path Programming   62
   3.6   Developing a Part Program   63
   3.7   Format Detail   66

3.8 ABSOLUTE PROGRAMMING 71
3.9 INCREMENTAL (RELATIVE) PROGRAMMING 73
3.10 ZERO SHIFT 75
3.11 SUMMARY 78

CHAPTER 4 INPUT MEDIA, FORMATS, AND PROGRAM TRANSFER — 81
4.1 INTRODUCTION 81
4.2 INPUT MEDIA 81
4.3 CODING SYSTEMS—RS–244 AND RS–358 85
4.4 METHODS OF ENTERING NUMBERS 91
4.5 PROGRAMMING FORMATS 92
4.6 PROGRAM TRANSFER METHODS 96
4.7 SUMMARY 101

CHAPTER 5 MATHEMATICS FOR NUMERICAL CONTROL PROGRAMMING — 103
5.1 INTRODUCTION 103
5.2 LAW OF SINES 103
5.3 LAW OF COSINES 109
5.4 POLAR NOTATION 112
5.5 CALCULATING CUTTER OFFSETS 114
5.6 SUMMARY 126

CHAPTER 6 LINEAR, CIRCULAR, AND HELICAL INTERPOLATION — 129
6.1 INTRODUCTION 129
6.2 LINEAR INTERPOLATION 130
6.3 CIRCULAR INTERPOLATION 133
6.4 HELICAL INTERPOLATION 137
6.5 SUMMARY 140

CHAPTER 7 PROGRAMMING CNC LATHES AND TURNING CENTERS — 143
7.1 INTRODUCTION 143
7.2 SETUP INFORMATION 143
7.3 TOOLING INFORMATION 145
7.4 PROGRAMMING AND OPERATING PROCEDURES 149
7.5 COMMON OPERATIONS PERFORMED 152
7.6 EXAMPLES 158
7.7 SUMMARY 159

CHAPTER 8 PROGRAMMING CNC MILLING MACHINES AND MACHINING CENTERS — 163
8.1 INTRODUCTION 163

- 8.2 Setup Information 164
- 8.3 Tooling Information 164
- 8.4 Common Operations Performed 167
- 8.5 Examples 180
- 8.6 Summary 183

Chapter 9 Other Machines Programmed Using Numerical Control 187
- 9.1 Introduction 187
- 9.2 Numerical Control Punching Machines 188
- 9.3 Electrical Discharge Machining 194
- 9.4 Flame-Cutting CNC Machines 196
- 9.5 Summary 200

Chapter 10 Repetitive Programming and Advanced Features 201
- 10.1 Introduction 201
- 10.2 Looping 202
- 10.3 Subroutines 205
- 10.4 Mirroring 208
- 10.5 Summary 211

Chapter 11 Computer Control in CNC Programming 215
- 11.1 Introduction 215
- 11.2 Aspects of the Computer in CNC 216
- 11.3 Computer Languages Available for CNC Programming 217
- 11.4 CAD/CAM 224
- 11.5 Distributive Numerical Control and Flexible Manufacturing 228
- 11.6 Summary 230

Appendices
- A Commonly Used National Codes 233
- B Useful Formulas and Tables 237
- C General Safety Procedures Relating to CNC Operations 241
- D Sample Part Programs 243
- E Glossary 249

Answers to Odd-Numbered Questions and Problems 257

Index 265

Introduction to Computer Numerical Control

Chapter Objectives

After studying this chapter, the student will be able to

- Recognize common varieties of NC and CNC machines.
- Identify the primary forms of input media.
- Discuss the general history of numerical control and computer numerical control.
- Define the terms *accuracy, reliability,* and *repeatability* as they relate to CNC machine technology.
- Identify and discuss the main types, uses, and advantages of CNC drive systems.
- Identify and discuss the primary types of feedback systems.
- List and explain the advantages and disadvantages of CNC systems.
- Recognize related applications of CNC technology.
- Identify the components of a CIM system.

1.1 Introduction

This chapter will provide a general overview of numerical and computer numerical control concepts related to the programming of machines used in the manufacturing industries. This chapter introduces many topics related to both **NC** (numerical control) and **CNC** (computer numerical control) technologies that you may encounter if you choose to pursue a career in one of the many manufacturing industries.

Subsequent chapters will deal with specific topics in greater depth. Throughout the text, study the examples given and try to determine the reasoning and motivation supporting them. Understanding the examples will be useful when you start writing programs and developing your own part programming style and technique.

Keep in mind that the primary objective of this text is to acquaint you with the technologies and procedures involved in numerical control and computer numerical control programming of machines. The purpose of this introductory chapter is to supply you with an understanding of the basic terminology and operating processes and procedures associated with NC/CNC machines so that we can build on this background. The principles given for CNC machines also apply to NC machines except where the difference is noted.

1.2 NUMERICAL CONTROL AND COMPUTER NUMERICAL CONTROL

The concept of numerical control began in the 1940s in response to the need for advanced manufacturing techniques to machine complex aircraft sections. Numerical control technology is simply the application of digital methods to control machines. Numerical control programming does not manufacture the part; it tells the machine how, when, and where to move to manufacture the part.

Numerical control programming is the actual physical and mental activity associated with designing and documenting a **part program** that will be used to manufacture a part. NC programming is often called **manual part programming** because it is performed without a computer. Numerical control programming performed using the computer may be termed computer-aided part programming (CAPP) or computer-aided manufacturing, discussed later.

Numerically controlled machines perform the same cutting and forming tasks used for decades in various manufacturing industries. The major difference and the principal advantage of NC equipment is the increased control of the cutting tool. Increased machine control allows for the manufacturing of parts that would be difficult or impossible to produce otherwise.

Coded part programs provide information used by the machine control unit (**MCU**) to control and position the cutting tool. The MCU is the brain of the NC machine. It functions much like the human brain in that it reads, interprets, and converts perceived input into appropriate movements. It also controls various accessories such as coolant, tool changes, and graphics. The MCU (also called the controller) converts the coded part program information into voltage or current pulses of varying frequency or magnitude, which are used to position and control the operation of the machine.

Most NC/CNC machines are able to store part programs in their memory. These machines memorize or store the program into memory when the program is first read into the machine. They can then call these programs from memory repeatedly, without having to read them again. This provides for faster operation when producing a number of identical parts. Machines without memory must read the part program a statement at a time and execute that statement before proceeding. Since they cannot store the program, machines without memory must reread the program each time a new part is produced. This slows operation considerably.

Strictly speaking, NC machines do not have on-board computer systems, although they may be equipped with buffer memory to store part programs prior to execution. MCUs equipped with buffer memory can remember an entire part program that is read into it. NC and CNC machines have two primary types of memory: (1) volatile RAM (random-access memory); and (2) nonvolatile ROM (read-only memory).

Part program input into the MCU is generally stored in RAM. RAM is able to store information while power is supplied to the MCU. Once all power is removed from the MCU, however, the information contained in RAM is generally lost. ROM is able to store the information read into it even after power is removed. Once information is written to ROM, it remains there until erased or overwritten. The MCU may be equipped with ROM that contains information concerning diagnostics, startup parameters, operating instructions, on-line help, and similar repeated functions. The advantage of storing these functions in ROM is that they only have to be programmed once and they are not lost when power is removed.

In addition to memory, the MCU also contains the hardware and software necessary to read, interpret, convert, and communicate the programmed statements in order to execute the proper machine functions. Keep in mind that the MCU is a separate but integral part of NC and CNC machines.

Computer numerical control developed in response to the need of manufacturing industries to produce higher quality parts with reduced lead times. But CNC machine manufacturers realized that just linking a computer to a production machine was not enough. To realize the full benefit of the computer, the machine should have its own on-board computer system.

CNC technology is based on the NC technology previously discussed, with the addition of on-board computer equipment to aid program processing. CNC part programs are programmed in the same fashion as NC part programs, using the added capabilities of the computer to read, store, edit, and process programmed information. These machines allow the operator to read program data, analyze the programmed information, optimize machining parameters, and edit the stored part programs. They also provide graphical capabilities, diagnostic procedures, and system troubleshooting. The addition of the computer enhances the advantages of NC technology by providing these extended capabilities as part of the basic machine functions.

NC and CNC machines may look similar and it is often hard to distinguish between them. Often the only clue that a machine has CNC capabilities is the graphics screen or cathode ray tube (**CRT**). In summary, the primary difference between NC and CNC machines is that CNC machines have on-board computer systems capable of extended functions such as graphics, diagnostics, and program editing.

The examples in this text center on the CNC equipment predominant in the field of manufacturing. This equipment includes milling machines and machining centers, and lathes and turning centers. Other machines you may encounter are NC punching machines, electrical discharge machines, and flame cutters. The fundamentals presented in this text apply to other NC and CNC applications as well.

Other applications of CNC technology include robotics, programmable controllers, automated drafting equipment, and programmable appliances. Although these machines are controlled numerically through the aid of a computer, they are not considered CNC equipment. They are considered types of automated manufacturing equipment that use similar technologies, but are only relatives to the CNC equipment we will be discussing. Do not confuse automated equipment with CNC equipment; they have similar advantages and objectives, but are not the same.

Who uses CNC machines? A wide variety of manufacturers and industries use NC and CNC machines. These range from large mass-production facilities to small job shops. Due to the wide variety of NC/CNC equipment used in industry, many different goods and services are produced. The aerospace industry is a good example. Because of the need for close tolerances and use of special alloys and exotic materials, many parts for the aerospace industry are machined using expensive, multifunction CNC machines. Because of the costs involved, many smaller production facilities may not be able to justify the expense of extended capabilities or graphics.

How much do CNC machines cost? They vary in cost according to the capabilities in number of axes (directions of programmable machine movement) controlled simultaneously, graphics capabilities, tooling capabilities, and other such related features. CNC machines may cost as little as $10,000 (for tabletop or instructional models) to as much as $2,000,000 or more, depending on the capabilities and accessories required. CNC machines come in a variety of standard designs or can be specially ordered from the manufacturer based on individual needs. Notable CNC equipment manufacturers include Bridgeport, Cincinnati-Milacron, Tsugami, Mazak, Mori-Seiki, and Hurco, among others. These and other companies can provide you with further information on the standard features and capabilities of CNC equipment. Fig. 1.1 illustrates some common CNC equipment.

FIGURE 1.1 COMMON CNC EQUIPMENT

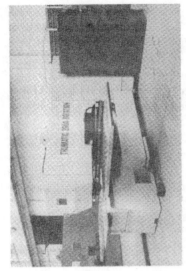

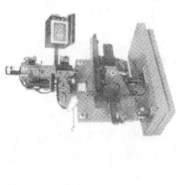

1.3 Part Programming and Input Media

In order for a CNC machine to produce a part, the programmer must write a representative part program. A part program consists of a coded set of alphanumeric characters and symbols whose purpose is to guide a numerically controlled machine through a set of predetermined movements to produce a part according to assigned specifications.

Let's break this definition down to simpler terms. A part program is a set of coded statements. The MCU has a specific language that it can interpret. For the MCU to recognize the part program, the statements must be written in the same language as the MCU. The Electronics Industries Association (EIA) and other organizations have standardized part-program coding systems (the language of the MCU). Information on coding systems is presented in Chapter 4. The programmer must code the part features into statements that the MCU can understand.

The part drawing provides the programmer with the necessary geometry and specifications to formulate the part program. Arcs, grooves, holes, and other part features are illustrated on the drawing, along with their locations. The part programmer then decides the proper order and sequence of operations to be performed. These are the predetermined movements of the machine based on the design specifications.

The success of the program depends on the experience and ability of the programmer to interpret and code the design specifications. A good part program requires careful planning and follows an efficient and logical flow to completion. Fig. 1.2 illustrates a basic part program.

```
N10 G0 G90 X-5 Y0 S1200 T1 M6
N20 X1 Z.1 F15
N30 G1 Z-.5
N40 Z.1
N50 G0 X2
N60 G1 Z-.5
N70 Z.1
N80 G0 X3
N90 G1 Z-.5
N100 Z.1
N110 G0 X-5 M2
```

The preceding program drilled three holes 0.5 in. deep spaced 1 in. apart. Chapter 3 provides further information about the meaning of the statement elements. For now, study the part drawing and part program to familiarize yourself with the format of each so that you can recognize them later.

FIGURE 1.2 ◻◻◻◻◻◻◻◻◻◻◻◻◻◻◻◻◻◻◻◻◻◻◻◻◻◻
SAMPLE PART PROGRAM

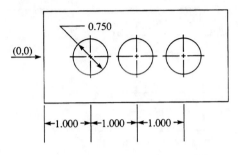

The part program in Fig. 1.2 could have been input in one of four ways: manual data input (**MDI**) using the machine's keypad; punched tape; magnetic media such as tape or floppy disks; or direct downloading through an RS–232 interface (such as in direct or distributive numerical control, discussed later in this chapter).

CNC machines typically contain an alphanumeric keypad with the characters and symbols necessary for writing part programs. The keypad also allows editing of a previously stored program. At this point, look at the MDI keypad in Fig. 1.3. Using the part program for Fig. 1.2, verify that the characters and symbols used in the part program are available on the keypad.

Punched tapes, as a primary form of program input media, are quickly becoming artifacts of the past. Due to the increased popularity and availability of the microcomputer, the use of the punched tape and tape reader has diminished. However, punched tapes are still used as a primary backup for part programs. Many users specify that their machines be equipped with tape readers in case the computer fails. Chapter 4 provides further discussion of the formats and procedures associated with punched tape.

Magnetic media such as magnetic tapes and floppy disks provide a means to store programs written on other devices, such as the computer. Floppy disks have become the storage media of choice due to their low cost and availability. Magnetic tapes resemble standard audiocassettes and can store several part programs. Of the two forms, floppy disks are the more popular programming medium.

In direct downloading operations, the MCU receives information, such as part programs, sent to it from an external source, such as a computer or disk reader via a physical connection. A hardwired connection, such as an interface cable, connects the external device and the MCU. The term RS–232 refers to an industry standard for data communications.

FIGURE 1.3
KEYPAD

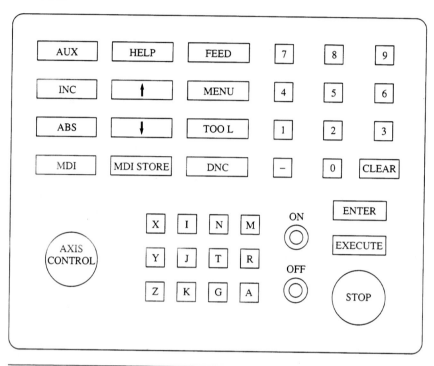

Machines equipped with standard ports or external connections may communicate through an interface. Most computers have at least one serial port for communicating with external devices such as printers, plotters, mice, and other peripherals. The serial port provides a link with the outside environment, in this case, the CNC machine. Fig. 1.4 illustrates the four principal input media presented in the previous discussion.

Two important CNC configurations discussed later in this text are **direct numerical control** and **distributive numerical control,** both abbreviated DNC. CNC machines can read and execute programs directly from memory. In direct numerical control, a central computer partially or completely controls the operation of one or more machines. This setup allows central control of the manufacturing system, although the increased cost of the coordinating computer and software makes direct control very expensive. In direct numerical control, each of the NC machines gets every command from the host. Direct numerical control has not been widely used in the U.S. Fig. 1.5 illustrates the direct numerical control process.

Distributive numerical control uses a network of computers to coordinate the operations of several machines. In fact, this network can con-

FIGURE 1.4
FOUR PRINCIPAL TYPES OF INPUT MEDIA

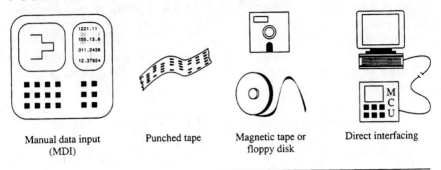

Manual data input (MDI)   Punched tape   Magnetic tape or floppy disk   Direct interfacing

FIGURE 1.5
DIRECT NUMERICAL CONTROL

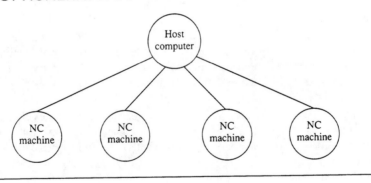

FIGURE 1.6
DISTRIBUTIVE NUMERICAL CONTROL

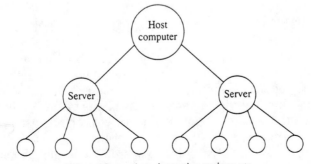

NC machines, robots, inspection stations, etc.

trol the entire manufacturing operation of a company. In distributive numerical control, each of the CNC machines is capable of storing all of the necessary commands for a single program or group of programs that have been received from the host. Distributive numerical control is widely implemented in the manufacturing industry and its use is growing in a great variety of applications. The distributive numerical control process is further examined in Chapter 11. For comparison with Fig. 1.5, Fig. 1.6 illustrates the distributive numerical control process.

1.4 HISTORY OF NUMERICAL CONTROL

Numerical control of machines developed in response to the growing problems that faced manufacturing industries before its development. These problems included a need to reduce lead times and inventory; increased market demand for greater options; and a reduction in the time it took to take a product from the development stage to actual production. Precursors of the modern CNC industry include the method devised by Joseph Jacquard in 1801 to control by punched cardboard instructions the patterns produced by textile looms. Another early example of numerical control is the player piano, which was controlled by a drum fitted with raised pins or a roll of paper through which air blew. Player pianos date back to the 1860s.

There is some dispute over who is responsible for the development of NC industrial technology. Many companies and organizations worked concurrently on the concept of numerical machine control during the 1940s. Important dates in the development of the modern numerical control industry include

1. **1947**—John C. Parsons of the Parsons Corporation, Traverse City, Michigan, manufacturers of helicopter rotor blades, could not produce templates for rotors fast enough. This dilemma led to Parsons' development of a method of coupling an early computer to a jig borer. Initially, Parsons used punched cards to code the necessary instructions for the Digitron System. This is generally considered to be the first application of NC technology.

2. **1949**—As the technology used in and intricacy of military equipment grew, the U.S. Air Materiel Command had been developing increasingly complex parts for its aircraft and missiles. To speed its production, the Air Materiel Command commissioned an Air Force study contract to Parsons to apply the numerical control system to its operations. Parsons subcontracted part of the study out to the Massachusetts Institute of Technology's Servomechanisms Laboratory.

3. **1951**—MIT took over the Air Materiel Command's contract and introduced the prototype for today's NC machines in 1952. Researchers at MIT coined the term *numerical control*.
4. **1955**—Seven companies exhibited tape-controlled contour milling machines at a national machine tool show. Each of these machines cost several hundred thousand dollars. In addition, these machines required trained mathematicians and powerful computers to produce the tapes necessary to run them. However, manufacturers began to realize the importance of numerical control and the potential for widespread application of the numerical control concept. Although these machines were programmed with computers, they were still numerically controlled through punched tapes and, therefore, NC machines.

After 1955, NC machine sales began to grow and their costs began to come down. This was due, in part, to the growing acceptance of NC and the continued miniaturization of electronic components required to program them.

From vacuum tubes to very large scale integrated circuits (VLSICs), electronic components shrunk significantly in size and cost. Both production and reliability increased, while necessary controller size and complexity decreased. Numerically controlled machines continued to impress manufacturers by performing operations that were previously thought impossible or impractical. In addition, these machines performed with increased accuracy and repeatability over conventional methods.

Once the concept of numerical control for machines gained acceptance from the manufacturing industry, NC machine sales skyrocketed. Along with increased sales came a demand for increased speed and reliability. The on-board computer provided the necessary tool for faster and more reliable CNC equipment. This helped manufacturers gain ground in the war to increase productivity and quality, a primary reason for the development of numerical control.

The NC machines produced in the 1960s relied on electronic hardware based on the digital circuit technology available at that time. CNC machines, primarily introduced in the 1970s, are less dependent on hardware and more dependent on software. They employ a minicomputer or microcomputer for machine tool control and reduce the need for supportive hardware. This trend away from hardware toward computer software has continued to bring increases in productivity, reliability, repeatability, and flexibility in the programming of CNC-produced parts.

In addition to advances in the hardware associated with CNC equipment, there have been significant advances made in the software used. Initially, programs were entered manually through punched paper tapes or keypunched through an alphanumeric keypad connected to the NC machine. Today's technology allows floppy disk program storage of programs entered via the computer. Programmers can now enter data and CNC programs using any of a variety of computer programs and languages.

Demands for increased production and accuracy have plagued manufacturers who have been concerned with the decline in the rate of productivity increase of U.S. workers since the early 1980s. This has led to increased levels of automation in U.S. manufacturing in an attempt to regain a competitive edge in the global marketplace. In addition, these demands have led to an increased reliance on computer software to program automated equipment and specifically CNC machines.

Three terms presented in the previous discussion are important to understanding numerical control technologies: **accuracy, repeatability,** and **reliability**. Take a minute to define each of these terms for yourself. Then compare your definitions with the following definitions.

Accuracy can be defined as the "trueness" of the manufactured part, or how close the part meets the specifications. An accurate part is free from error, being manufactured within all specified tolerances. Time and experience are necessary on a manually controlled machine with which the operator is very familiar to produce parts that maintain a tolerance of $+/- 0.001$ in. (0.025 mm). This contrasts with CNC machines currently available that are capable of maintaining tolerances of $+/- 0.000010$ in. (0.00025 mm) or better. Both types of machines rely on the ability of the operator to maintain the setup and cutting tools.

The following factors influence the accuracy of the CNC machine:

1. Proper maintenance at the proper time
2. Proper cutting loads
3. Proper foundation setup according to manufacturer's specifications
4. Environmental conditions
5. The material being machined
6. The condition and selection of cutting tools
7. The condition and selection of tool holders
8. Programming experience of the operator

Repeatability is the ability of the machine to produce similar parts every time. The same factors that affect accuracy affect the repeatability of the machine. Typically, the machine's repeatability should be one-half of the actual **positioning tolerance** of the machine. Positioning tolerance is the ability of the machine to locate repeatedly at the same point in space.

The reliability of the machine relates to its availability to produce parts and avoid downtime. A reliable machine is free from excessive downtime or excessive maintenance. One term related to reliability is Mean Time Between Failure (MTBF). This is the calculated average length of time the machine is in service between breakdowns. The larger the MTBF, the more production the manufacturer can gain from the machine and the lower the cost of maintaining the level of production.

1.5 DRIVES AND CONTROL LOOPS

The drive motors that control the axes of machine movement on NC and CNC equipment come in four basic types: stepper motors, DC servos, AC servos, and hydraulic servos. Stepper motors have a discrete number of positions or steps that they can travel. When the controller receives the programmed instruction to move in an axis, it sends a number of signals to the stepper motor. The number of signals or pulses that the controller sends to the stepper motor controls the amount of rotation of the motor. In addition, stepper motors use the frequency of the pulses to control the speed of the programmed motions, and, therefore, the feed rate of the machine. Stepper motors are cheaper to control and produce. They require less associated hardware and are generally used on lower-cost machines.

AC and DC servos are used on small- to medium-sized CNC contouring machines. They are variable speed motors that rotate in response to the applied voltage. DC servo motors operate on varying voltage magnitude. AC servo motors are controlled by varying the voltage frequency to control speed. The MCU is able to control machine positioning by metering the amount and frequency of voltage sent to the servo. AC servos can develop more power than DC servos. Larger machines use servos rather than stepper motors.

Hydraulic servos are also variable speed motors. Because they are hydraulic, they are able to produce more power than electric servos. Large-capacity machines use hydraulic servos for axis control. Hydraulic servos are controlled by an electronic or pneumatic control system that meters the amount of fluid to the drive.

There are two types of control systems on NC machines: **open-loop** and **closed-loop**. The type of control loop used determines the overall accuracy potential for the machine. Normally, open-loop systems use stepper motors, while closed-loop systems use AC, DC, or hydraulic servos.

The open-loop control system does not provide positioning feedback to the MCU. The MCU sends the necessary positioning information to the axis drive motor. The drive motor then processes the information and the MCU sends the next positioning instruction for processing. There is no provision for positioning error detection or correction.

Advances in the manufacture and control of stepper motors have made the open-loop process more accurate and dependable. However, there is still room for error. The advantage of the stepper motor is that it does not require the additional hardware and electronics needed to provide positioning feedback. They are less expensive and more compact than closed-loop systems. The disadvantage is the lack of positioning error detection and correction.

In the closed-loop system, feedback is available to detect when an error has occurred. The MCU compares the current position of the driven axis or axes with the programmed position. When the two coincide, the MCU stops the drives and starts to process the next instruction. If the MCU detects an error and it cannot correct the error, it may stop program execution at that point and give an error message (depending on the severity of the error). The principal advantage of the closed-loop system is it can provide the MCU with position information so that the MCU can detect and correct for most positioning errors.

1.6 Advantages and Disadvantages of Numerical Control

Since its introduction, NC technology has found many applications. CNC machines are manufactured in many standard sizes and varieties. Each machine is available with a variety of options and controllers. The one thing that all of these machines have in common is that they are numerically controlled. Of the various numerically controlled machines available, common applications include CNC milling machines and machining centers, lathes and turning centers, punches, testing and inspection equipment, electrical discharge machines (EDM), flame cutters, and precision grinders. In general, all of these machines have the following objectives:

1. Perform functions impossible or impractical by other methods
2. Increase the accuracy and repeatability of parts
3. Reduce production costs
4. Increase production levels

With these objectives in mind, consider the following lists of advantages and disadvantages of numerical control.

Advantages

1. Increased productivity
2. Improvements in the control of manufacturing
3. Increased quality control
4. Increased repeatability in parts
5. Reduced material handling (robotics, palletizers, etc.)
6. Faster setup times

7. Reduced tool/fixture storage and costs
8. Reduced parts inventory
9. Increased accuracy
10. Increased flexibility and response to design changes
11. Creation of new jobs

Disadvantages
1. Increased maintenance
2. High initial financial investment required
3. Higher per-hour operating cost
4. Need for retraining or replacement of personnel

This is not a complete list of the advantages and disadvantages of numerical control machines. These lists are a general guide to the benefits and problems that may occur through the implementation of CNC technology. They are also a guide to the decision whether to implement these technologies.

1.7 FMS, CAD/CAM, AND CIM

Although the development of numerical control technology began in the aerospace industry, numerical control methods apply to a variety of manufacturing situations. Advances in electronics have allowed further increases in the accuracy, repeatability, and reliability of NC machines. The cost of NC machines has also come down, allowing smaller shops to purchase and use them. The applications of numerical control technology in modern manufacturing industries are widespread, yet new applications and advances in older applications are constantly being made. With these advances, there is an increased demand for competent programmers with the knowledge and ambition required to tackle these advances.

Current applications of CNC technology rely heavily on microcomputers. The manufacturing industry has evolved from long, continuous production lines down to smaller work cells based on the flexible manufacturing system concept. Flexible manufacturing systems (FMS) can be more readily rearranged to meet different production requirements. FMS, combined with automated assembly and inspection facilities, has become standard in today's manufacturing industry. This system is supervised by the highly sophisticated computer-integrated manufacturing (CIM) systems developed in response to the desired change from hardware-dependent to software-reliant systems previously discussed. These CIM systems control and manage the production flow from conception to finished product.

It is beyond the scope of this text to adequately cover all of the concepts related to FMS and CIM. However, a few basic terms will be discussed here and in greater depth in Chapter 11.

When several CNC machines and a robot are merged into a single unit to produce one part or group of parts with similar geometry, they are said to be a manufacturing cell or work cell. Each piece of equipment is controlled by a computer, which communicates with a main cell computer. The main cell computer coordinates the activities of the machines within the cell including loading, unloading, tool changes, part changes, and similar activities. The software used to control these operations is also capable of handling machine breakdowns, tool breakage, out-of-stock conditions, and other situations. The manufacturing cell is the basic unit of a flexible manufacturing system. Each cell computer communicates with a central computer in the system. The central computer controls the operations of the facility from rough stock to finished part.

FMS uses **computer-aided design or drafting/computer-aided manufacturing** systems **(CAD/CAM)** to integrate the design, documentation, and manufacturing components, which reduces the total production cycle time. CAD/CAM systems are able to design parts around specific design criteria based on the operating parameters of the system. For example, one criterion may be the part's suitability for CNC machining compared to EDM. Another criterion may be the suitability of the material for robotic assembly versus manual assembly.

Once the design is created, CAM allows the programmer to trace a tool path for the part. CAM systems generally allow for the selection of tooling parameters and the optimum tool path for greatest efficiency. The CAM output can then be sent to the CNC machines, robots, and other components of the manufacturing cell. The CAD/CAM computer must have access to the data associated with the part or part group and cell capabilities in order to be truly effective.

CIM systems coordinate the efforts of computers in each aspect of manufacturing, including product design, production planning, part production, assembly, testing and inspection, and the flow of materials and parts throughout the facility. Ultimately, CIM systems include all facets of the industry including accounting, bookkeeping, purchasing, inventory, scheduling, and all other aspects of the business. The central computer in a CIM Shop Floor Management system (Area Controller) controls and monitors the tasks performed by the member computers, regulating these activities based on overall management strategies.

The advantages of the CIM system stem from the ability of the central computer to coordinate activities in response to changes in the production environment. CIM systems also reduce the amount of direct labor required to produce parts and reduce required parts inventories.

In summary, the basic components of a CIM system are the manufacturing cell, the FMS, CAD/CAM, and the management system. These components work in harmony to increase the overall productivity and efficiency of today's manufacturing enterprises.

Throughout this text, some of the many applications of CNC technology will be discussed. While studying the examples, try to determine how CNC technology is applied to the example and how related technologies are used. This will provide you with a better insight into the control and functions of CNC technologies.

1.8 SUMMARY

A numerically controlled machine produces parts based on a coded set of instructions fed into it by tape or computer. NC machines are not equipped with on-board computer systems, although they may contain buffer memory, which stores part programs. The memory may be either RAM (random-access memory) or ROM (read-only memory). RAM is volatile in that the information stored in it is lost when power is removed. ROM is nonvolatile. CNC machines have on-board computer systems that are able to read and execute programs directly from memory. Direct and distributive numerical control processes involve the control of machines by external computer systems through interfaces. In direct numerical control operations, each machine under control receives each instruction from the host computer. In distributive numerical control, the machines under control can store one or more programs downloaded from the host.

There are four types of drives used on CNC equipment: stepper motors, DC servos, AC servos, and hydraulic servos. There are two types of control loops on CNC machines: open-loop and closed-loop. Open-loop systems do not have feedback capabilities. Closed-loop systems can detect and correct positioning errors, while open-loop systems cannot. There are four ways to input CNC programs: MDI, punched tape, magnetic media, and direct interfacing. The advantages and disadvantages of NC are factors in the decision to adopt numerical control technologies.

The basic components of a CIM system are the manufacturing cell, the FMS, CAD/CAM, and the management system. These components work in harmony to increase the overall productivity and efficiency of today's manufacturing enterprises.

QUESTIONS AND PROBLEMS

1. List the NC/CNC machines used at your facility.
2. Define numerical control as you understand it.
3. Describe in your own words a numerical control part program as you understand it.

4. What is your definition of the difference between NC and CNC?
5. Discuss the history of NC and CNC technologies. Try to relate social and technological events with the progress of these technologies.
6. List and describe the four drive systems found on CNC equipment.
7. Differentiate between open- and closed-loop systems.
8. In your own words, define the following terms: MDI, NC, CNC, RS–232 interfacing, direct numerical control, and distributive numerical control.
9. List four ways in which a program may be input into an NC machine.
10. What are the objectives of numerical control?
11. List the advantages and disadvantages of numerical control technology.
12. Choose a machining task you are familiar with and describe how CNC programming differs from conventional methods.
13. Describe the following terms as they relate to numerical control methods: work cell, FMS, CIM, CAD, CAM.

TOOLING FEATURES FOR MILLING AND TURNING MACHINES

CHAPTER OBJECTIVES

After studying this chapter, the student will be able to

- Identify common insert terminology and classifications related to CNC tooling.
- Differentiate between nine different cutting tool materials.
- Identify common cutting tool material coatings.
- List common cutting tool properties.
- Define and understand the use of qualified and preset tooling.
- Define tool radius or tool diameter compensation.
- Recognize the types and advantages of various CNC tool changers.
- Calculate proper feed rates and speeds for CNC operations.
- Identify the purposes of adaptive control systems.
- Discuss the advantages of adaptive control systems related to CNC tooling and programming.

2.1 INTRODUCTION

This chapter will describe the cutting tools commonly used in CNC operations and certain parameters associated with this tooling. In addition, some of the many tool changer options available for machining and turning centers are described. Because correct speeds and feed rates optimize production rates and maximize tool life, the CNC programmer should be well acquainted with tool materials and the calculation of speeds and feed rates for various materials. Parts of this chapter may be a review if you are familiar with insert composition and terminology. If so, please review the material briefly and move on to material you are less familiar with.

2.2 CUTTING SPEEDS, SPINDLE SPEEDS, AND FEED RATES

Three important terms related to cutting tool performance are cutting speed, spindle speed, and feed rate. **Cutting speed** refers to the speed at which a point on the edge of the tool or cutter travels in relation to the workpiece. This is measured in **surface feet per minute (FPM** or **SFPM)** or meters per minute (m/min). The cutting speed indicates that either the tool is moving past the workpiece or the workpiece is moving past the tool at a specified distance per minute. The cutting speed for a material is a predetermined value or range of values based on research and practical experience. Recommended cutting speeds for several common materials are given in Appendix A. More complete information on recommended cutting speeds for various materials and cutter profiles may be found in machinist's handbooks, common engineering references, or from the manufacturer of the material.

Spindle or **chuck speeds** are normally specified in **revolutions per minute (RPM)**. Spindle speed and cutting speed are related, but distinctly different. Spindle or chuck speed refers to the number of revolutions made by the tool (on milling machines) or workpiece (on turning machines) per minute, independent of the tool selected or workpiece dimensions. Normally, the cutting speed specified for a material is necessary to calculate the desired spindle speed. Spindle speeds may be programmed as constant RPM or constant surface speeds. Constant RPM maintains a constant chuck or spindle speed while cutting. Constant surface speeds vary the spindle RPM to maintain the desired surface feet per minute cutting rate (cutting speed). **Constant surface speed (CSS)** is normally used when programming turning operations. The following shows both calculations, cutting speed and spindle speed.

$$\text{spindle RPM} = \frac{\text{cutting speed} \cdot 12 \text{ (in./ft)}}{\text{diameter (in.)} \cdot \pi} \quad (2.1)$$

where the cutting speed in surface feet per minute is found in the tables and the diameter is taken from the part or cutter.

An approximation of this formula is

$$\text{spindle RPM} = \frac{\text{cutting speed} \cdot 4}{\text{diameter}} \quad (2.2)$$

$$\text{cutting speed (FPM)} = \frac{\text{RPM} \cdot \text{diameter (in.)} \cdot \pi}{12 \text{ (in./ft)}} \quad (2.3)$$

where RPM is the spindle speed in revolutions per minute and the diameter is the diameter of the cutter in inches.

SECTION 2.2 CUTTING SPEEDS, SPINDLE SPEEDS, AND FEED RATES

An approximation of this formula is

$$\text{cutting speed (FPM)} = \frac{\text{RPM} \cdot \text{diameter (in.)}}{4} \quad (2.4)$$

where 4 is an approximation of $12/\pi$.

In the metric system,

$$\text{cutting speed (m/min)} = \frac{\text{RPM} \cdot \text{diameter (mm)} \cdot \pi}{1000 \text{ (mm/m)}} \quad (2.5)$$

An approximation of this formula is

$$\text{cutting speed (m/min)} = \frac{\text{RPM} \cdot \text{diameter (mm)}}{300} \quad (2.6)$$

where 300 is an approximation of $1000/\pi$.

EXAMPLES

Calculate the milling machine spindle speed required for a cutting speed of 100 FPM using a 0.5 in.-HSS end mill. Using the formula and a tool diameter of 0.5 in.,

$$\text{spindle RPM} = \frac{\text{cutting speed} \cdot 12 \text{ (in./ft)}}{\text{diameter (in.)} \cdot \pi}$$

$$\text{RPM} = \frac{100 \text{ FPM} \cdot 12}{0.5 \text{ in.} \cdot \pi} = 764 \text{ RPM}$$

Based on a spindle speed of 1000 RPM, what would the cutting speed in FPM be for the end mill in the previous example? Using the formula,

$$\text{cutting speed (FPM)} = \frac{\text{RPM} \cdot \text{diameter (in.)} \cdot \pi}{12 \text{ (in./ft)}}$$

$$\text{FPM} = \frac{1000 \text{ RPM} \cdot 0.5 \text{ in.} \cdot \pi}{12 \text{ (in./ft)}} = 130.9 \text{ FPM}$$

Calculate the spindle speed required on a lathe for a cutting speed of 100 FPM and a 1.25-in. diameter piece of stock.

$$\text{spindle RPM} = \frac{\text{cutting speed} \cdot 12 \text{ (in./ft)}}{\text{diameter (in.)} \cdot \pi}$$

(For the lathe, we use the stock diameter rather than the tool diameter as in the previous example.)

$$\text{RPM} = \frac{100 \text{ FPM} \cdot 12 \text{ (in./ft)}}{1.25 \text{ in.} \cdot \pi} = 306 \text{ RPM}$$

If the required cutting speed were 200 FPM, what would the effect be on the spindle speed? Using the formula,

$$\text{chuck speed (RPM)} = \frac{\text{CS (FPM)} \cdot 12}{\text{diameter (in.)} \cdot \pi}$$

$$\text{RPM} = \frac{200 \text{ FPM} \cdot 12}{1.25 \text{ in.} \cdot \pi} = 611 \text{ RPM}$$

When using the constant surface speed option, the spindle or chuck speed will increase or decrease according to decreasing or increasing stock diameter, respectively. This maintains the constant programmed cutting speed of the workpiece in relation to the cutting tool. Constant RPM and constant surface speed are programmed using appropriate codes. Normally, constant RPM is implied, unless programmed otherwise.

Feed rate is the rate at which the tool or cutter moves into the workpiece. Many variables influence the value of the feed rate, including the spindle or chuck speed, chip load per tooth, number of teeth on the cutter, and material. Feed rates for some common materials are in Appendix A. More complete information may be found in machinist's handbooks, common engineering references, or from the manufacturer of the material.

Programmed feed rates are in **inches per minute (IPM)** or **inches per revolution (IPR)**. IPM feeds are time-based and independent of spindle or chuck speed. IPR feeds directly relate to spindle or chuck speed, feeding the tool or cutter at a designated speed per spindle or chuck revolution. The operator may override the programmed feed rate through the feed rate override control, if the machine is so equipped. This allows the operator to adjust the feed rate from 0 to 200% (typically) of the programmed value. The following provides feed rate calculations for common CNC operations.

$$\text{milling feed rate (IPM)} = \text{RPM} \cdot T \cdot N \tag{2.7}$$

or

$$\text{feed (mm/min)} = \text{RPM} \cdot T \text{ (mm/rev)} \cdot N \tag{2.8}$$

where RPM is the spindle speed in revolutions per minute, T is the chip load per tooth in inches per revolution or millimeters per revolution (found in Appendix A), and N is the number of teeth on the cutter.

$$\text{lathe feed rate (IPM)} = \text{RPM} \cdot \text{IPR} \quad (2.9)$$

or

$$\text{feed (mm/min)} = \text{RPM} \cdot \text{(mm/rev)} \quad (2.10)$$

where IPR is the feed rate in inches per revolution.

This formula is used for drills, reamers, countersinks, and lathe programming.

$$\text{lathe feed rate (IPR)} = \frac{\text{IPM}}{\text{RPM}} \quad (2.11)$$

or

$$\text{lathe feed rate (mm/rev)} = \frac{\text{mm/min}}{\text{RPM}} \quad (2.12)$$

where IPM is the feed rate in inches per minute and RPM is the spindle speed.

EXAMPLES

Calculate the milling feed rate for 1200 RPM using a four-flute cutter with a chip load of 0.002 chip per tooth. Using the formula,

$$\text{milling feed rate (IPM)} = \text{RPM} \cdot T \cdot N$$

$$\text{IPM} = 1200 \text{ RPM} \cdot 0.002 \text{ chip per tooth} \cdot 4 = 9.6 \text{ IPM}$$

Calculate the IPR feed rate for a lathe using a feed rate of 12 IPM running at 2000 RPM. Using the formula,

$$\text{lathe feed rate (IPR)} = \frac{\text{IPM}}{\text{RPM}}$$

$$\text{feed rate (IPR)} = \frac{12 \text{ IPM}}{2000 \text{ RPM}} = 0.006 \text{ IPR}$$

Calculate the IPM feed rate for a lathe using a feed rate of 0.020 IPR at 1000 RPM. Using the formula,

$$\text{lathe feed rate (IPM)} = \text{RPM} \cdot \text{IPR}$$

$$\text{IPM} = 1000 \text{ RPM} \cdot 0.020 \text{ IPR} = 20 \text{ IPM}$$

Following are further examples of calculating the spindle speed and feed rate required for various operations. Calculate the spindle speed in RPM and feed rate in inches per minute for a 0.25 in.-HSS drill running at 100 FPM cutting speed.

$$\text{spindle RPM} = \frac{\text{cutting speed} \cdot 12 \text{ (in./ft)}}{\text{diameter (in.)} \cdot \pi} = 1528 \text{ RPM}$$

$$\text{feed rate (IPM)} = (\text{RPM}) \cdot (\text{IPR}) \text{ [from Appendix A]}$$
$$= 1528 \text{ RPM} \cdot 0.004 \text{ IPR} = 6 \text{ IPM}$$

Calculate the spindle speed and feed rate in IPM for a 0.5 in.-HSS four-flute milling cutter running at 100 FPM cutting speed with a chip load of 0.002 chip per tooth.

$$\text{spindle RPM} = \frac{\text{cutting speed} \cdot 12 \text{ (in./ft)}}{\text{diameter (in.)} \cdot \pi} = 764 \text{ RPM}$$

$$\text{feed rate (IPM)} = 764 \text{ RPM} \cdot 0.002 \text{ IPR} \cdot 4 = 6 \text{ IPM}$$

Proper speeds and feed rates help ensure maximum productivity and tool life. Too high a speed will dull or burn up a tool or result in a poor surface finish. Too low a speed will break the tool and waste machine time. Too high a feed rate may break the tool or produce a rough surface finish. Too low a feed rate wastes machine time. The following factors affect the selection of cutting feed rates and speeds:

1. The surface finish required
2. Whether the cut is continuous or interrupted
3. The amount, type, and efficiency of the cutting fluid used
4. Specified tolerances
5. Depth of part features
6. Material of the workpiece
7. Heat treatment of the workpiece
8. Tool cost and maximum tool life desired

The following general statements may be made concerning cutting speeds and feed rates. For heavier cuts, slower speeds are programmed. Lighter cuts are generally programmed at higher speeds. Cutting RPM and feed rates are specified by the programmer in the part program. Programming these values correctly depends on the experience and knowledge of the programmer. It is important to note that the speeds and feed rates calculated are starting points and not absolutes. Generally, RPM and IPM are specified in CNC milling applications, while CSS and IPR are specified in CNC turning applications. These values are adjusted based on actual machining experience and tool performance.

2.3 Cutting Tool Materials

Cutting tool materials, such as carbon steels and high speed steel, used in the past met the needs of industry at that time. Today, these materials are still used but are not suitable for tougher and harder materials such as the new space-age metals and alloys. The development of tougher and harder materials and alloys and the higher production rates at which industry must run have demanded increases in cutting tool technology and materials. The materials discussed in this section are high carbon steel, high speed steel (HSS), cast alloys, cemented carbides, coated carbides, ceramics, cermets, diamond, and cubic boron nitride.

High carbon steel cutting tools are generally restricted to lower cutting speeds and temperatures. These tools tend to soften at cutting speeds above 50 FPM in mild steels or temperatures above 250°C (482°F). Therefore, they are used for softer materials such as wood, aluminum, brass, copper, and magnesium.

High speed steel cutting tools are alloy steels, which contain primarily tungsten and chromium. They may also contain small percentages of cobalt, molybdenum, and vanadium. HSS cutting tools are relatively inexpensive and tough. Since it is a relatively tough material, HSS cutting tools can be used at higher cutting speeds (up to 100 FPM in mild steels) and higher tool contact temperatures (up to 550°C or 1022°F). HSS is sometimes used to produce special-shaped tools, such as those for boring, for lathes and turning centers. It is also commonly used in the manufacture of milling cutters and drills.

Recently, high speed steels have been coated with various materials to increase their wear resistance and lower the tool's coefficient of friction. Materials such as titanium nitride (TiN) have been used as a thin coating (1 to 2 μm) over a HSS core. The high speed steel core provides a ductile, shock-resistant foundation, while the coating provides additional wear resistance and lower friction than uncoated cutting tools.

Cast alloys are nonferrous (contain no iron) alloys that are cast into their final shape. Cast alloy cutting tools are softer than HSS cutting tools, but retain their shape at higher temperatures. They generally consist of carbon, cobalt, chromium, and tungsten. The use of cast alloy cutting tools is generally reserved for machining cast iron, malleable iron, and hard bronzes. They can be used at cutting speeds up to 200 FPM and temperatures up to 650°C (1202°F).

A carbide is a chemical compound composed of carbon and a metal. When referring to cutting tools, the term carbide generally refers to cemented carbides. Cemented carbides are composed of tantalum carbide, titanium carbide, or tungsten carbide and cobalt in various combinations. Cemented carbides are manufactured by compressing the component metal powders and sintering the material at temperatures of about 1400°C (2552°F). Sintering is a process by which heat is used to weld individual particles together without melting them. Cemented carbides have high hardness values even at temperatures reaching 1200°C (2192°F). They can therefore be used at higher cutting speeds than HSS or cast alloy cutting tools (about 400 FPM in mild steels). Cemented carbides are not as tough as HSS or cast alloy cutting tools and cannot be reshaped once they have been sintered. Therefore, cemented carbides are generally available in insert form, whether brazed or clamped onto a tool holder. Clamped-on inserts are thrown away once all of the cutting edges have been used. A new insert must be brazed on in brazed applications. Tungsten carbide cutting tools are the strongest and offer the highest wear resistance, but suffer rapid deterioration when machining steels. Therefore, tantalum carbide and titanium carbide are added to improve the properties of straight tungsten carbide. These materials are designated as tungsten-tantalum carbide or tungsten-titanium carbide.

Another method of achieving the toughness of tungsten carbide while achieving the superior wear resistance of titanium carbide is to coat a core of tungsten carbide with a thin layer (5 to 8 µm) of harder material. These coated carbides offer the best cutting speeds that current carbide cutting tool technology can offer. The materials used to coat carbides are generally titanium carbide, titanium dioxide, and titanium nitride. Recently, other materials have also been used with success: aluminum oxide (Al_2O_3), hafnium nitride (HfN), and zirconium nitride (ZrN). Experiments have also been made with multiple layers, where more than one coating layer is applied, for example, a coating of aluminum oxide covered by a layer of titanium nitride. The use of refractory materials (such as aluminum oxide) as coatings helps to inhibit heat buildup at the point of tool contact and extends tool life. Average tool life for coated carbide varieties is generally five times that for uncoated materials. Higher cutting speeds and extended tool life have made coated carbide a popular choice in milling and turning applications.

Ceramics or cemented oxide tool materials are made from sintered aluminum oxide mixed with various boron and silicon nitride powders, which are compressed together at pressures over 4000 PSI (pounds per square inch) and sintered at temperatures reaching 1700°C (3092°F). Some manufacturers add titanium, magnesium, or chromium oxides in small amounts to enhance the cutting properties. The manufacturing process used to produce these ceramics enhances the density and hardness of these cutting tools. Cutting tools made of cemented oxides can be used at two to three times the cutting speeds of tungsten carbide cutting tools. They are very hard, yet are brittle and require that shock and vibration be minimized. Cemented oxide cutting tools are available only in insert form and are most often used to machine relatively hard materials at high speeds.

Cermets are combinations of ceramics and metals. One common type of cermet cutting tool is made from 70% aluminum oxide and 30% titanium carbide. The high strength, high shear resistance of the aluminum oxide ceramic is combined with the toughness and thermal shock resistance of the titanium carbide metal. These cermet materials are generally tougher, more fracture-resistant, but softer than the ceramics. This allows them to be used at speeds much higher than those for straight carbides and to perform interrupted cuts. One other combination involves ceramic-ceramic composites. For example, silicon carbide whiskers may be imbedded in aluminum oxide to help disperse the cutting forces. This increases the mechanical properties of the ceramic-ceramic composite over that of the straight cermets.

Diamond cutting tools are made by compressing small particles of diamond (either natural or synthetic) under high pressure at elevated temperature. Natural diamonds are the hardest materials known. However, natural diamonds have faults and nonuniform strength; this makes them unsatisfactory for most manufacturing operations. Therefore, they are ground and compressed to give diamond tooling improved strength and durability with only a slight reduction in overall strength. Synthetic diamond tooling made up of super-hard elements is commonly used in CNC applications. This synthetic diamond tooling is known as polycrystalline diamond or PCD. Diamond tooling resists the compression forces produced during machining twice as well as tungsten carbide and offers little thermal expansion. It is often used in precision machining and fine finish operations. However, diamond tooling is restricted to nonferrous, nonmetallic materials. When machining ferrous materials, a chemical reaction takes place at high cutting temperatures and can cause diamond to break down to its original graphite form. Diamond and PCD tooling are classified as "carbon-seeking," which renders them unsuitable for machining ferrous or metallic materials.

Another very hard cutting tool material is cubic boron nitride or CBN. CBN is less expensive than diamond, but much more expensive than carbide. However, CBN tooling will last over 50 times longer than carbide tooling.

The tooling used to machine a part must be harder than the workpiece. Some very hard materials require diamond or CBN tooling. The advantages of using these cutting tools are the fine finishes possible and very high cutting speeds (2000 to 2500 FPM).

Some general guidelines concerning tooling are

1. Inspect all tools before they are used. Dull or broken tools should be replaced immediately. Check to make sure tools/inserts are tight in the tool holder and the correct toolholder is being used.
2. Select the proper tool to perform the operation. Do not attempt to drill with an end mill or mill with a twist drill. Use the tool properly and select the proper size tool to do the job safely and efficiently.
3. Maintain the proper cutting feeds and speeds for the tool used. Proper speeds and feeds maximize efficiency and tool life. Improper speeds

and feeds waste time and money in decreased tool life, slow production times, and tool changes.
4. Become acquainted with the machine's capabilities. Often, machines are used inefficiently because programmers do not know they could perform a simpler operation on the machine or have been performing an operation incorrectly, causing excessive tool wear or tool breakage.
5. When using machine tools, program the largest depth of cut and highest feed rate safely possible. This reduces cycle time and has a negligible effect on tool life.
6. Use coolants to extend tool life, clear chips, and reduce heat buildup in the workpiece.

2.4 Carbide Insert Terminology and Applications

Consider for a moment the three forms in which cutting tools are commonly available: solid tooling, brazed tips, and indexable inserts. Drills are a good example of solid tooling. They are available in many styles and materials such as high carbon steel, high speed steel, and coated or uncoated carbide. However, once they are worn they must be resharpened or thrown away. A second form of cutting tools is the brazed tip, in which a single-edged cutting tool is bonded to a medium-carbon steel tool holder. Standard lathe tooling, for example, is available in brazed-tip form, in which a carbide cutting tool is brazed onto a standard tool holder. Drills, reamers, milling cutters, and turning tools are standard brazed-tip applications. Brazed tools can be resharpened by grinding a number of times before replacement is necessary. However, it takes time to mount and replace or sharpen brazed tips. The third form of cutting tools is the indexable insert. Carbides and cemented carbides as inserts for special tool holders provide the widest applications in manufacturing. These inserts are available for many applications including milling and turning. Figure 2.1 details the insert/tool holder combination.

Consider the following when selecting the proper carbide insert for an application:

1. Operating conditions
2. Carbide grade
3. Insert size and shape
4. Tool style
5. Tool nose radius

FIGURE 2.1 ☐☐☐☐☐☐☐☐☐☐☐☐☐☐☐☐☐☐☐☐☐☐☐☐☐☐☐
INSERT/TOOL HOLDER COMBINATION

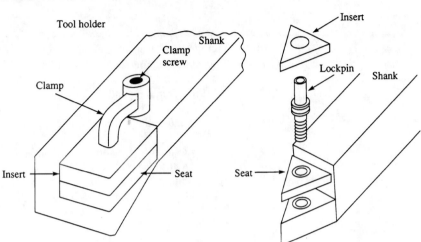

Three operating conditions are used to determine the metal removal rate for an application: cutting speed, feed rate, and depth of cut. Cutting speed has the most effect on tool life. The depth of cut is affected by the size and thickness of the insert and the hardness of the material being machined. In general, harder materials necessitate lower feed rates, lower cutting speeds, and smaller depths of cut. The depth of cut is also limited by the strength and thickness of the insert, horsepower of the machine, and the amount of material to be removed.

There is a wide range of carbide grades, insert shapes, and tool holder designs available. However, the American National Standards Institute (ANSI) and the International Standards Organization (ISO) have helped standardize these grades, shapes, and designs by introducing standards for their manufacture. Most manufacturers will state whether they comply with these ANSI/ISO codes.

Wear resistance and toughness are used to grade carbide inserts. Generally, as wear resistance increases, toughness decreases. ANSI and ISO standards differ in carbide grading systems. Manufacturers can typically supply both sets of information. The ANSI allows manufacturers to create their own carbide grades and classifications, which are explained in their catalogs. The ISO has developed a standard grading system that groups carbides according to application. This system employs a set of letters (P, M, and K) and a number. The letters are also assigned a corresponding color code (P = blue, M = yellow, and K = red). Table 2.1 illustrates the ISO grading system.

TABLE 2.1
ISO CARBIDE GRADING SYSTEM

Category Color	Material Usage	Code	Application
P (Blue)	Ferrous Metals	P01	Vibration-free finish turning and boring.
		P10	Turning, threading, and milling. Light roughing and finishing.
		P20	Turning, threading, and milling. Average cutting conditions.
		P30	Turning and milling with heavy feeds and shocks.
		P40	Turning and milling castings and stainless steels under unfavorable conditions.
		P50	Turning and milling under unfavorable conditions requiring very tough carbide.
M (Yellow)	Ferrous Metals	M10	Finish turning alloy steels with light feeds and moderate speeds.
		M20	Turning and milling of alloy steels. Universal machining grade.
	Nonferrous Metals	M30	Roughing of alloys that are difficult to machine.
		M40	Turning, parting off, and roughing of mild steel, nonferrous metals, and alloys.
K (Red)	Ferrous Metals	K01	Finishing hard materials and abrasive polymers. Wear-resistant.
	Nonferrous Metals	K10	Turning, milling, and drilling of malleable cast iron and nonferrous alloys.
	Nonmetallic Materials	K20	Turning, milling, and boring of nonferrous alloys requiring tough carbide.
		K30	Rough turning and milling under unfavorable conditions.
		K40	Turning and milling of materials under unfavorable conditions.

As an example, use the ISO grading system to determine the carbide grade best suited for finishing stainless steels. From the chart in Table 2.1, the selection is M10.

Cemented carbides are also graded according to application. These grades were assigned by the Cemented Carbide Producers Association (CCPA). For example, the grades assigned by the CCPA for uncoated carbides are

Grade	Application	Material
C–1	Roughing Cuts	Cast iron, nonferrous materials
C–2	General Purpose	Cast iron, nonferrous materials

FIGURE 2.2
RAKE ANGLE

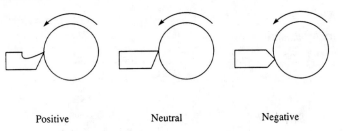

Grade	Application	Material
C-3	Light Finishing	Cast iron, nonferrous materials
C-4	Precision Boring	Cast iron, nonferrous materials
C-5	Roughing Cuts	Steel
C-6	General Purpose	Steel
C-7	Finishing Cuts	Steel
C-8	Precision Boring	Steel

where the hardest of the nonferrous and cast iron grades is C-4 and the hardest of the steel grades is C-8. These grades were assigned based on experience. Grade selection may require modification based on actual performance data. These are broad categories, which are meant as application guides and not as absolute mandates.

Indexable inserts are available in a number of standard sizes and shapes. They are made to be reusable in that they have a number of cutting edges. After all of the cutting edges have been used, the insert is

FIGURE 2.3
INSERT SHAPES

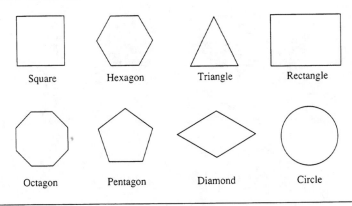

32   CHAPTER 2   TOOLING FEATURES FOR MILLING AND TURNING MACHINES

FIGURE 2.4
ANSI Indexable Inserts Identification System

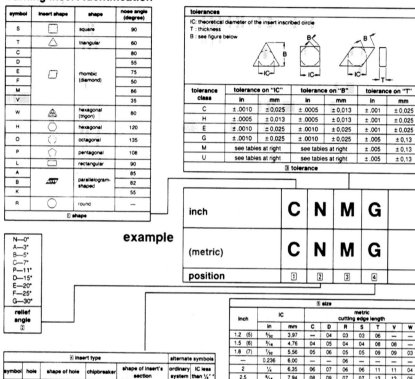

SECTION 2.4 CARBIDE INSERT TERMINOLOGY AND APPLICATIONS

Technical Data

± tolerance on "IC"

IC		class "M" tolerance						class "U" tolerance	
		shapes S, T, C, R & W		shape "D"		shape "V"		shapes "S, T & C"	
inch	metric	inch	mm	inch	mm	inch	mm	inch	mm
5/32	3,97			—	—	—	—	—	—
3/16	4,76			—	—	—	—		
7/32	5,56	.002	0,05						
1/4	6,35			.002	0,05	.002	0,05	.003	0,06
5/16	7,94								
3/8	9,52								
7/16	11,11								
1/2	12,70	.003	0,06	.003	0,06	.003	0,06	.005	0,13
9/16	14,29								
5/8	15,88								
11/16	17,46	.004	0,10	.004	0,10	.004	0,10	.007	0,18
3/4	19,05								
7/8	22,22	.005	0,13	—	—	—	—	.010	0,25
1	25,40								
1 1/4	22,22	.006	0,15	—	—	—	—		

± tolerance on "B"

IC		class "M" tolerance						class "U" tolerance	
		shapes S, T, C, R & W		shape "D"		shape "V"		shapes "S, T & C"	
inch	metric	inch	mm	inch	mm	inch	mm	inch	mm
5/32	3,97			—	—	—	—	—	—
3/16	4,76			—	—	—	—		
7/32	5,56	.003	0,06					.005	0,13
1/4	6,35			.004	0,11	—	—		
5/16	7,94					.007	0,18		
3/8	9,52							—	—
7/16	11,11							—	—
1/2	12,70	.005	0,13	.006	0,15	.010	0,25	.008	0,20
9/16	14,29							—	—
5/8	15,88							—	—
11/16	17,46	.006	0,15	.007	0,18	—	—	.011	0,27
3/4	19,05							—	—
7/8	22,22							—	—
1	25,40	.007	0,18	—	—	—	—	.015	0,38
1 1/4	22,22	.008	0,20	—	—	—	—		

```
 —    4    3    2    ☐    ☐    ☐
      12   04   08   ☐    ☐    ☐
      [5]  [6]  [7]  [8]  [9]  [10]
```

[5] thickness

thickness		symbol	
in	mm	inch	metric
1/32	0,79	0.5 (1)	—
1/16	1,59	1 (2)	01
5/64	1,98	1.2	T1
3/32	2,38	1.5 (3)	02
1/8	3,18	2	03
5/32	3,97	2.5	T3
3/16	4,76	3	04
7/32	5,56	3.5	05
1/4	6,35	4	06
5/16	7,94	5	07
3/8	9,52	6	09
7/16	11,11	7	11
1/2	12,70	8	12

NOTE: Inch sizes in parenthesis for "alternate sizes" D or E (under 1/4 inch IC).

[8] hand of insert (optional)

R | L

[7] corner radius

corner radius		symbol	
in	mm	inch	metric
.004	0,1	0	01
.008	0,2	0.5	02
1/64	0,4	1	04
1/32	0,8	2	08
3/64	1,2	3	12
1/16	1,6	4	16
5/64	2,0	5	20
3/32	2,4	6	24
7/64	2,8	7	28
1/8	3,2	8	32
round insert (inch)		—	00
round insert (metric)		—	M0

[9] & [10] cutting edge condition or chip control features (optional)

- T — negative land
- K — light feed chip control, double sided Kenloc insert
- M — heavy feed chip control, deep floor Kenloc
- N — narrow land Kentrol insert with chip control on one side
- W — heavy-duty chip control, wide land Kenloc insert one side
- J — polished to 4-microinch AA (rake face only)
- UF — ultra-fine finishing

See Technical Section for additional conditions and chip control features.

thrown away. They cannot be resharpened. The standard insert shapes include round, square, and triangular.

Round inserts provide the greatest strength, as well as the largest number of cutting edges. However, they are limited to applications not affected by their radius, generally straight turning operations.

Square inserts are not as strong and provide fewer cutting edges than round inserts. They are stronger than triangular inserts and provide a greater range of application than round inserts. They provide a maximum of eight possible cutting edges (four on top, four on bottom, depending on the rake).

Tool rake ranges from positive to negative and affects the formation of the chip and the surface finish. Zero and negative rake tools are stronger and provide longer life than positive rake tooling. Figure 2.2 on p. 31 illustrates the three types of rake angles.

Triangular inserts provide the most versatility. However, they are weaker than round or square inserts and offer fewer cutting edges (three on top, three on bottom, depending on the rake). Operations such as combination turning and facing and tracing are commonly performed with triangular inserts.

Diamond-shaped inserts are also available. These inserts have two acute and two obtuse cutting edges on each side (top and bottom) although, generally, only the acute cutting edges are used. The acute angles on these inserts are commonly 35°, 55°, and 80°. The angular shape allows for a wider range of part profiles and part features to be cut with a single cutting tool. Fig. 2.3 on p. 31 illustrates the common insert shapes.

In addition to shape, size is an important factor in insert selection. Generally, the smallest insert capable of producing the desired depth of cut at the programmed feed rate is selected. The length of the cutting edges should be 1½ times the length that the insert contacts the workpiece. For example, if the cutting tool contact is 0.5 in., then a 0.75 in. face length should be selected.

A third factor, thickness, also affects insert selection. Insert thickness affects insert strength, the thicker, the stronger. Feed rate and depth of cut influence the desired insert thickness. The carbide grade is also a factor in choosing insert thickness. Again, actual insert performance is the best guide in choosing an insert. For common roughing applications insert thicknesses of 0.25 or 0.375 in. are generally adequate. Fig. 2.4 on pp. 32 and 33 shows the ANSI indexable inserts identification system, which is commonly used in selecting indexable inserts.

Here is an example of how to use the ANSI system to grade a CNMG–632 insert.

C   Insert shape: 80° diamond

N   Relief angle: 0° relief angle (where relief is provided by the tool holder)

SECTION 2.4 CARBIDE INSERT TERMINOLOGY AND APPLICATIONS

M   Insert tolerance: ±0.002 or ±0.004 on inscribed circle (I.C.), ±0.005 thickness

G   Type: with hole and chipbreaker

6   Size: ¾ in. I.C.

3   Thickness: 3/16 in.

2   Cutting point radius between flats: 1/32 in. radius

Figure 2.5 depicts the relief angle, insert shape, inscribed circle, and cutting point radius features described in the previous example. Chipbreakers are designed to curl the chip produced by the insert back into the workpiece and then break it off to produce the desired type of chip. Chipbreakers may be formed as part of the insert itself or clamped into the tool holder with the insert.

The choice of a tool style depends on a particular operation and the associated operation parameters. Choosing a tool holder/insert combination requires familiarity with the machine being used and the operation being performed. Once a choice of design has been made, the carbide manufacturers and ANSI have devised a system for identifying tool holders used with carbide inserts. This grading system is used to specify the tool holder geometry. Fig. 2.6 illustrates the ANSI tool holder identification system commonly used by manufacturers and suppliers.

Here is how this system is used to determine the specifications for a CANN–12 tool holder:

C   Insert shape: 80° diamond

A   Holder style: straight shank with 0° side cutting edge angle

N   Rake: negative

FIGURE 2.5
INSERT IDENTIFICATION

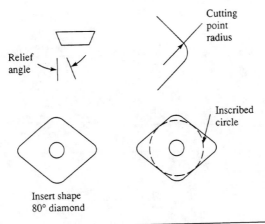

Relief angle

Cutting point radius

Inscribed circle

Insert shape
80° diamond

CHAPTER 2 TOOLING FEATURES FOR MILLING AND TURNING MACHINES

**FIGURE 2.6
ANSI TOOL HOLDER IDENTIFICATION SYSTEM**

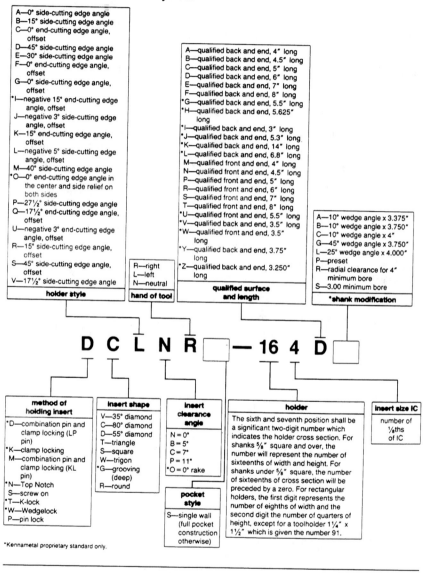

N   Hand of tool: neutral
12   Holder: number of 16ths of width and height on square shank

The tool nose radius is important in terms of tool strength, surface finish, and desired part features. Consider the feed rate, depth of cut, and workpiece material when selecting a proper tool nose radius. In general, larger radii are stronger and produce better surface finishes. Select the largest radius that does not produce tool chatter.

2.5 Tooling Systems

Generally, in order to produce a part, several different cutting tools are used. The way in which these tools are stored and held in position while machining is known as a tooling system. The type of tooling system used affects the overall machining time and productivity of the machining operation.

Machines that are not equipped with automatic tool changers require that the operator manually change tools. In manual tool changes, the operator removes the old tool and replaces it with the new tool, tightens the new tool, and restarts the machine. This process is inefficient and time-consuming.

To reduce the time required for tool changing, quick-release chucks with special tool holder combinations are often used. This process still requires the operator to stop and restart the machine, which is inefficient in both time and productivity.

Automatic tool changers allow tool changing without the intervention of the operator. Machining and turning centers usually employ automatic tool changers with drums, turrets, magazines, or carrousels. Automatic tool changing ability is commonly used to distinguish the more versatile CNC machining and turning centers from CNC milling machines and lathes.

Several options are available in automatic tool changers. Typically, the tool changer grips the tool, placing it in a position that aligns it with the spindle. The tool and tool holder combination is then inserted into the spindle and locked into position by the draw stud. The tool changer then moves out of the way to some park location until the next tool is called by the program.

Two terms related to automatic tool changes are random and sequential tool selection. Random tool selection allows the selection of tool positions at random. Tool position refers to the location of tools within the tool magazine. A tool magazine is an indexable tool storage facility that stores tools until they are needed. When a tool is called in the part program, the tool magazine will index to the proper tool, where it is retrieved by the tool handling device. The location of the tool magazine in relation to the spindle varies among the tool change systems used. Each tool has a unique tool location within the magazine, which uniquely identifies the tool to the MCU. As tools are called, the MCU indexes the tool magazine to the proper tool location. Tools do not have to be located in the magazine in the order they are called in the program; they may be located in any position within the tool magazine.

In sequential tool selection, however, the tool must be located in the tool magazine in the order in which they are called by the program. Random tool selection is more versatile. Sequential tool selection, in some circumstances, may be faster (less time is required for indexing or searching for the next tool), but may require additional setup time where tooling requirements often vary. Many modern machines allow indexing of the tool magazine during program execution, in effect, searching for the next tool while the previous operating step is performed. This eliminates the time advantage of sequential tool selection.

There are five basic types of automatic tool changers commonly used on CNC machines: indexable turret, 180° rotation, pivot insertion, two-axis sweep (an axis may be thought of as a programmed direction of machine movement), and spindle direct. The procedure for calling a tool change is basically the same for all of these systems. The process used to change tools varies between the five types.

A turret head is among the oldest of automatic tool changing methods. It contains a number of spindles attached to the machine head. The proper tools are located in the spindles prior to program execution. As tools are called, the turret rotates or *indexes* to the proper position, where the tool has been loaded. The main disadvantage of turret heads is the limited number of tool spindles available. This limits the number of tools called during the program, unless the operator stops the program during execution and substitutes new tool/holder combinations for some of the used combinations. The turret head tool changer is often used on CNC punching machines. Fig. 2.7 shows a typical turret head arrangement.

The simplest of the automatic tool changers is the 180° rotation. When given the tool change command, the spindle travels to the preset tool change position. The tool turret then indexes to the proper tool position, where the tool changer rotates to engage both the current tool and the tool to be loaded. The draw stud is then removed from the current tool and both tools are removed from their positions. The tool changer then rotates 180°, swapping the current tool for the requested tool. As the tools are being swapped, the tool turret rotates to the old tool position so that the old tool may be replaced in the tool turret. The tool changer then rotates back to its fixed position, out of the way of the spindle, and remains there until another tool change is called. The advantage of the 180° rotation is the simplicity of the tool change system, minimizing movement while maintaining adequate tool change speed. Fig. 2.8 illustrates the 180° rotation method of tool changing.

FIGURE 2.7
TURRET HEAD TOOL CHANGER

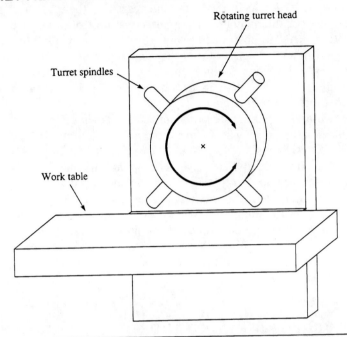

FIGURE 2.8
180° ROTATION

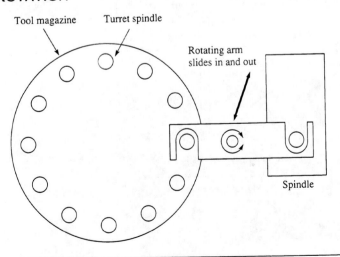

Pivot insertions, an adaptation of the 180° rotation, has been the most common type of automatic tool change system. The same procedure is used in pivot insertion as in 180° rotation except that the tool turret pivots out of the way of the machine. The main advantage of pivot insertion is that the tool turret is positioned away from the hazard of flying chips, coolant, and other foreign material that accompanies material removal. However, pivot insertion requires more movement and therefore has a greater tool change time than 180° rotation. Fig. 2.9 depicts the pivot insertion method of tool changing.

The two-axis sweep system is similar to pivot insertion in that the tool changer is located away from the spindle, often at the side of the machine. The two-axis sweep moves in two axes to change tools rather than pivoting. Its main disadvantage is the great amount of tool handling necessary with two-axis movement and the increased amount of tool-change time required. It has the advantage of providing maximum tool protection. Fig. 2.10 details the two-axis sweep method of tool changing.

The spindle direct system moves the tool turret directly to the spindle or the spindle may move directly to the tool turret, depending on machine size. Maximum efficiency dictates which method is used. On larger ma-

FIGURE 2.9 ☐☐☐☐☐☐☐☐☐☐☐☐☐☐☐☐☐☐☐☐☐☐☐☐☐☐☐☐
PIVOT INSERTION

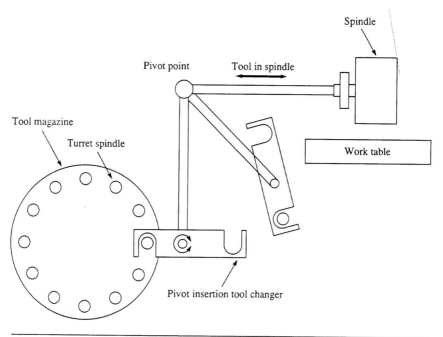

FIGURE 2.10
TWO-AXIS SWEEP

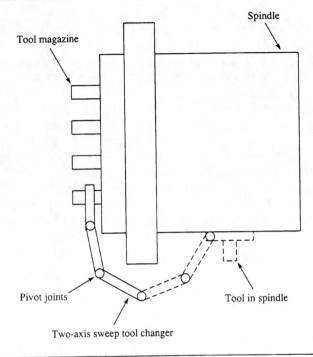

FIGURE 2.11
SPINDLE DIRECT

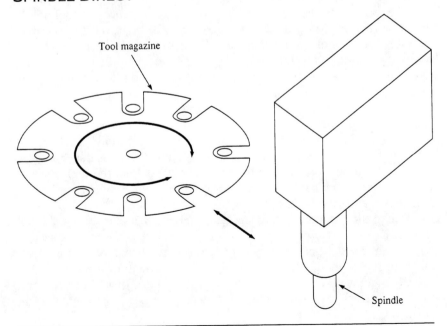

chines, the tool turret may be too large for it to practically move to the spindle; in such cases the spindle must move to the tool turret. This system cuts down on tool handling and tool change time, allowing direct tool changes without intermediate steps. Fig. 2.11 illustrates the spindle direct method of tool changing.

To summarize, tool selection in automatic tool changing may be sequential or random, depending on the machine's capability and the tool magazine setup used by the operator. In order to maintain efficiency, tools are located at the shortest rotation distance. Since some turrets index back to a set tool position when parked, frequently used tools are located close to the park position of the turret. This park position is often referred to as the home position for the tool turret or tool magazine. Comments reflecting tool selection and position may be placed in the program manuscript. Remember, these are the basic tool changing methods used on CNC machines; specific machine designs may vary according to manufacturer.

2.6 Tool Lengths and Tool Length Offsets

CNC programming requires some method of informing the machine how far the tool projects from the spindle or tool holder. To supply this information, the tool length or tool preset length is measured.

The tool length (as shown in Fig. 2.12) is the distance the tool holder and tool combination project beyond some fixed point on the machine, typically the face of the machine spindle or turret. Measuring each tool manually is tedious work and makes tool replacement difficult. In order to maintain the accuracy of programmed parts run on the machine, each new tool would have to project the exact same distance from the machine as the last tool. This is not practical for the wide variety of cutting tools used in CNC operations.

Manual tool length measurement is, however, widely practiced on CNC machines. But the success of the tooling operation is dependent on the knowledge and experience of the person measuring the tool. Several mechanical and optical tool length setup devices are available to help in manually measuring tools.

Due to the variation in tool lengths, some method must be used to tell the MCU how to compensate for these differences. Programmable tool registers allow the operator to enter the tool number, tool length **offset,** and tool radius for a number of tools. This information is stored in the MCU's memory. When a particular tool is called up within a program, the MCU refers to this information to determine how much offset to al-

FIGURE 2.12
(A) TOOL LENGTH AND TOOL LENGTH OFFSET.
(B) TOOL OFFSETS FOR TURNING MACHINES

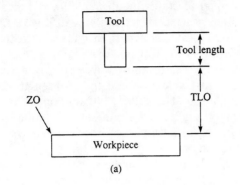

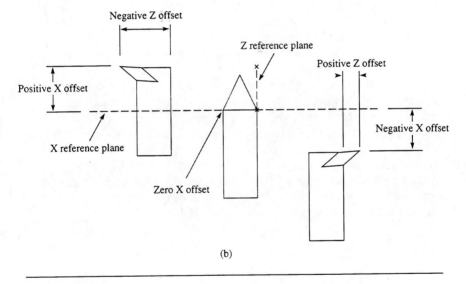

low. An example of common tooling information entered into a programmable tool register is

Tool #	Tool Length Offset	Radius
1	0.500	0.250
2	1.118	0.500
3	0.379	1.125
4	0.553	0.750
5	1.013	0.375

Other information may be included, such as tool length instead of offset. In this case, the MCU calculates the offset or uses the tool length directly. Feed rates and spindle speeds may also be included on some machines.

A *tool length offset* (TLO) is the distance from the cutting edge of a tool to the desired part surface when the tool is fully retracted. The tool length offset for a tool is not the length of the tool. Refer to Fig. 2.12 to help visualize the difference in tool length and tool length offset. The purpose of tool length offsets is to allow for clearance of clamps, fixtures, and other features. Most machines call the tool length offset for a tool automatically when the tool is called, but this operation is not always automatic. Some machines require the offset to be called separately. The operator must enter the offset figures corresponding to the appropriate tools before running the program.

Based on the differences in length, an arbitrary minimum offset is established and the difference between tool lengths, starting with the longest tool, added to the minimum offset. The longest tool generally has the shortest TLO. Another method of setting the TLOs involves bringing the tool in contact with the part surface (used on knee and column type milling machines, for example). This procedure begins with inserting the longest tool in the spindle and raising the workpiece until the tip of the tool touches the surface. The workpiece is then lowered the desired offset to clear clamps, fixtures, and other fixed features of the machine setup. Once this distance is set, the TLOs are entered into the MCU. The programmer may enter TLOs into the program (if known) or the operator can enter them during machine setup. This is only one method of entering TLOs.

The tool lengths may be measured manually and the difference between each succeeding tool added to a minimum tool length offset. Consider the following example.

EXAMPLE

If T1 had a length of 4 in., T2, 2 in., T3, 3 in., and T4, 5 in., with the minimum TLO set at 0.5 inch, the following calculations could be made to determine the offsets:

Tool #	Tool Length	TLO	Calculation
T1	4.000	1.500	(5.000 − 4.000) + 0.5
T2	2.000	3.500	(3.000 − 2.000) + 2.500
T3	3.000	2.500	(4.000 − 3.000) + 1.500
T4	5.000	0.500	(specified minimum)

The specified minimum TLO is assigned to the longest tool, T4. The next longest tool length (T1) is then subtracted from the length of T4. This value is added to the TLO for T4, and this sum becomes the TLO for T1. The third longest tool length (T3) is then subtracted from the second longest tool length (T1) and added to the TLO for T1. This sum becomes the TLO value for T3. This continues until all of the TLOs have been calculated. (See Fig. 2.13.)

The MCU interprets the TLO for each tool as the distance the tool extends from the machine each time a new tool is loaded. Proper setting of TLOs is important in maintaining proper part geometry. If incorrect offset values are programmed, they may: (1) cause the tool to follow the programmed machine movements without contacting the workpiece (if too small); or (2) cut too deep into the workpiece (if too large). Very large errors in TLOs may cause the tool to crash into the workpiece. Setting TLOs is an important procedure and strongly affects the accuracy of part features.

Correct tool length offset information is also important for CNC turning machines. These are referred to as the X and Z offset values, since the tool length and tool holder/insert combination may vary in both the X and Z axes. The X and Z offset values may be entered manually or in the part program, similar to the procedure described for CNC milling machines. Offset values for CNC turning machines, as for CNC milling machines, are the distances that the tool or insert projects from a specified reference point in both the X and Z axes.

Qualified and preset tooling can reduce tool change and tool setting times. The position of the cutting edges on qualified tools is guaranteed to within close limits of accuracy. The cutting edge position tolerance is

FIGURE 2.13 ☐☐☐☐☐☐☐☐☐☐☐☐☐☐☐☐☐☐☐☐☐☐☐☐☐☐
CALCULATING TLOS

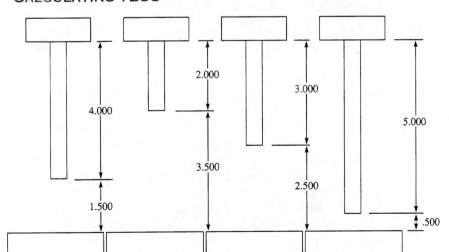

given with respect to specified data on the tool holder. Qualified tooling tolerances are generally within +/− 0.08 mm (+/− 0.0005 in.). Tool qualification is aptly applied to insert/tool holder combinations where precise dimensions and geometry are known. This reduces the need for measuring tools and calculating TLOs, as well as the time it takes to preset tooling. Qualified tooling reduces the amount of time that the machine stands idle while the operator measures tool lengths and calculates and enters tool offsets.

Presetting tools also reduces the amount of time required for tool changes. Tool presetting is often performed in the tool room with special apparatus. Presetting requires that the exact machine data be duplicated on a special fixture. Once mounted, precision positioning tables, digital readouts, and profile projection equipment are used to gain the desired positional accuracy of the cutting edge. Although tool presetting requires more time than qualified tooling, it still provides a significant reduction in the downtime required for manually measuring, entering, and adjusting for differences in tool lengths.

Both qualified tooling and preset tooling are available for turning and milling machines. Where specifications demand precise tool positioning, the programmer may wish to specify that qualified or preset tools be used. Qualified and preset tooling also reduce the chance of setup errors or tool dimensioning errors. If qualified or preset tooling is specified, the programmer should document this fact on the setup sheet submitted to the operator with the program.

2.7 TOOL RADIUS OR TOOL DIAMETER COMPENSATION

Most modern CNC machines offer some form of tool radius or tool diameter compensation. This allows for: (1) using different sized tooling than was originally programmed; (2) varying the size of part features cut in the workpiece; (3) making a series of cuts, as in roughing and finishing passes; (4) adjusting for the cutting edge radius; and (5) adjusting for tool wear.

Positive compensation values reflect oversize tooling or the amount of stock left on the material for finishing, while negative compensation values reflect undersize tooling or undercutting. When using tool diameter compensation, compensation values are included in the program with the required preparatory (G) codes. (G codes tell the MCU the manner in which you want the axes to move.) G codes are discussed in Chapter 3.

Once given the appropriate compensation code, the tool will change to the compensated position on the next programmed move. Canceling compensation causes the tool to change from the compensated position to an uncompensated position on the next programmed move. Briefly, the

SECTION 2.7 TOOL RADIUS OR TOOL DIAMETER COMPENSATION   47

G41 code initializes tool compensation to the left, G42 initializes tool compensation to the right, and G40 cancels tool compensation. To determine the direction of compensation, imagine yourself in the position of the tool facing the last direction of tool travel. Tool compensation to the left is on your left and tool compensation to the right is on your right. Fig. 2.14 shows the effects of tool diameter compensation on turning and milling machines.

An advantage of tool radius compensation is that it allows the use of oversize or undersize tooling, which means that the same basic program can be used to produce parts that vary in size. It may also be used to adjust for tool wear and allows for a series of cuts to be made along the same part profile. Compensation may be made within the program or manually entered into the MCU. Some controllers require additional code words when using tool radius compensation: the H code for programming tool length and the D code for programming tool diameters.

When tool radius compensation is programmed, the tool will move to the compensated position on the next programmed movement. The tool will ramp to and from the compensated position when turned on and off, respectively. These are noncutting moves necessary to correctly position the tool.

In the case of CNC turning machines, the tool nose radius (TNR) compensation value is taken into account. TNR compensation values may be entered manually or through the part program with codes and addresses similar to those on CNC milling machines. Different insert

FIGURE 2.14 □□□□□□□□□□□□□□□□□□□□□□□□□□
TOOL RADIUS OR TOOL DIAMETER COMPENSATION

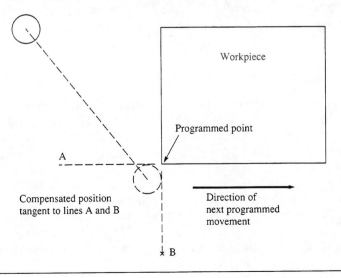

FIGURE 2.15 ▫▫▫▫▫▫▫▫▫▫▫▫▫▫▫▫▫▫▫▫▫▫▫▫▫▫▫▫
TNR COMPENSATION

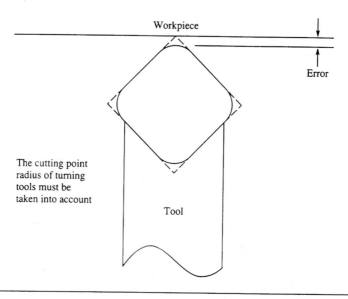

shapes, sizes, and configurations will have different tool nose radii. Therefore, each different tool may have a unique TNR compensation value. Insert specifications include the TNR value for a particular insert or cutting tool. These values and the X and Z offset values are the three components of the TLO values for CNC turning machines. Each value must be measured, calculated, and entered into the MCU in order to maintain proper part geometry. Figure 2.15 illustrates the reason for using TNR compensation.

2.8 ADAPTIVE CONTROL

One other concept related to tooling systems is **adaptive control.** Adaptive control allows the MCU to adjust machining parameters based on operating conditions. This is not an operator-controlled function. Adaptive control is the automatic response of the machine based on preprogrammed machining variables.

Adaptive control is not the same as feed override control. Adaptive control involves sensing variations associated with the process and optimizing the machining parameters based on this feedback. Horsepower and torque are two commonly monitored variables in adaptive control systems. The main difference between the two is that the operator manually controls the feed rate override control. Adaptive controls require additional circuitry to provide feedback to the MCU.

Adaptive control systems were developed to provide real-time adjustment of machining parameters. In many instances, the part programmer programs conservative values in an effort to preserve the life of the tool. This slows down machine production. In adaptive control, machine parameters can be adjusted automatically, which provides maximum production rates while preserving tool life.

Adaptive control systems are of two general types: (1) adaptive control with optimization (ACO); and (2) adaptive control with constraints (ACC). Practically all adaptive control systems used for machining processes are of the ACC type; development of the ACO type continues. However, there are major problems associated with defining realistic indexes of performance, and the lack of suitable sensors to reliably measure parameters makes ACO systems impractical. In ACC systems, machining parameters are maximized within a region bounded by process and system constraints or limits.

Adaptive control systems allow for 20 to 80% increases in production rates compared with conventional machining, depending on the properties of the material being machined and the complexity of the workpiece. In addition, the programmer is saved the time and effort of calculating and programming operating parameters such as speeds and feed rates, since these are automatically controlled by the ACC. This savings becomes significant when programming complex parts.

Adaptive control is a control technology related to CNC programming. For further information concerning the processes and procedures involved in adaptive control, refer to the many references available on the subject of machine control systems and automated systems control.

2.9 SUMMARY

The term *cutting speed* refers to the speed at which a point on the edge of the tool or cutter travels in relation to the workpiece, measured in surface feet per minute (FPM or SFPM) or meters per minute (m/min). The cutting speed indicates that either the tool is moving past the workpiece or the workpiece is moving past the tool at a specified distance per minute. The cutting speed for a material is a predetermined value or range of values based on research and practical experience.

Spindle or chuck speeds are normally specified in revolutions per minute (RPM). Spindle speed and cutting speed are related but distinctly different. Spindle or chuck speed refers to the number of revolutions made by the tool (milling machines) or workpiece (turning machines) per minute, independent of the tool selected or workpiece dimensions. Normally, the cutting speed specified for a material is necessary to calculate the desired spindle speed.

Feed rate is the rate at which the tool or cutter moves into the workpiece. Many variables influence the value of the feed rate, including the spindle or chuck speed, chip load per tooth, number of teeth on the cutter, and material. Correct cutting speeds and feed rates must be researched and calculated for each new material machined. It is important to program correct speeds and feeds to help ensure maximum efficiency and accuracy in machining. Factors that affect the selection of cutting feed rates and speeds include

1. The surface finish required
2. Whether the cut is continuous or interrupted
3. The amount, type, and efficiency of the cutting fluid used
4. Specified tolerances
5. Depth of part features
6. Material of the workpiece
7. Heat treatment of the workpiece
8. Tool cost and maximum tool life desired

Feed rates may be based on the spindle speed as in./revolution or independent of the speed as in./minute. Generally, RPM and IPM are specified in CNC milling applications, while CSS and IPR are specified and CNC turning applications.

Cutting tool materials have evolved in response to the needs of modern manufacturing industries. Older cutting tool materials were found to be inadequate for working with the new materials and higher production rates required by modern manufacturing industries. The cutting tool materials discussed in this chapter include high carbon steel, high speed steel (HSS), cast alloys, cemented carbides, coated carbides, ceramics, cermets, and diamond.

Three basic types of cutting tools are solid tooling, brazed tip, and insert. Indexable inserts are widely used in the manufacturing industry. They are indexable because as a cutting edge becomes worn, the insert is moved to a new position, where a new edge is brought in contact with the work.

In addition to cutting tool materials, there is a wide range of carbide grades, insert shapes, and tool holder designs available. However, the American National Standards Institute (ANSI) and the International Standards Organization (ISO) have helped standardize these grades, shapes, and designs by introducing standards for their manufacture. Most manufacturers state whether they comply with these ANSI/ISO codes.

Carbide cutting tools come in coated and uncoated varieties. Coated carbides are carbide cores over which is deposited a thin layer of coating.

The coating contains a harder material that provides the cutting tool or insert with the enhanced properties of the coating while maintaining the advantages of the carbide core. Coated carbides offer the best cutting speeds that current carbide cutting tool technology can offer. Materials used to coat carbides are generally titanium carbide, titanium dioxide, and titanium nitride. Recently, other materials, such as aluminum oxide (Al_2O_3), hafnium nitride (HfN), and zirconium nitride (ZrN), have been used with success. The use of refractory materials (such as aluminum oxide) as coatings helps to inhibit heat buildup at the point of tool contact and extends tool life. Average tool life for coated carbide varieties is generally five times that for uncoated materials. Higher cutting speeds and extended tool life have made coated carbide a popular choice in milling and turning applications.

Generally, in order to produce a part several different cutting tools are used. The way in which these tools are stored and held in position while machining is known as a tooling system. The type of tooling system used affects the overall machining time and productivity of the machining operation.

Two basic types of tooling systems are manual and automatic tool changers. Manual tool changes are performed by the machine operator. Automatic tool changers use automated equipment to select and replace cutting tools called by the part program. Automatic tool change systems are based on the ability of the machine to store and retrieve the cutting tools necessary to cut the programmed workpiece. These tools are stored in tool magazines, carrousels, turrets, and other similar holding devices. These devices are capable of holding anywhere from 10 to 100 or more cutting tools.

Automatic tool changers are capable of calling tools in random or sequential order. In random tool selection, tools may be called in any order without a specific pattern. Sequential tool selection specifies that tools are selected in the order in which they occur in the tool changer. For example, the first tool called by the part program must be in the first tool position, the second tool called must be in the second tool position, and so forth.

After tool changes, automatic tool changers may return to a home position. This home position is often a particular tool position such as Tool #1. In order to reduce the amount of time it takes to change tools, frequently selected tools should be located close to the home position.

Most modern CNC machines offer some form of tool radius or tool diameter compensation. Tool diameter compensation is a useful feature that allows for: (1) using different sized tooling than was originally programmed; (2) varying the size of part features cut in the workpiece; (3) making a series of cuts, as in roughing and finishing passes; (4) adjusting for cutting edge radii; and (5) adjusting for tool wear. In addition, the programming of tool radius compensation allows the programming of part surfaces and not the tool centerline on the mill or the center of the tool nose radius on the lathe. This allows for the maintenance of part geometry with different tool combinations.

Because tool lengths vary and must be measured before executing the program, tool lengths or tool length offsets must be entered into the MCU in order to maintain part geometry during program execution. One method of determining the tool length offset (TLO) for a number of tools involves measuring the length of each tool independently and assigning a specified minimum TLO to the longest tool. Each subsequent shorter tool will have an offset greater than the specified minimum by the difference between its length and the longest tool's length. These values (tool number, length, offset, and radius) may be entered into the MCU.

Adaptive control systems are of two basic categories: (1) adaptive control with optimization (ACO); and (2) adaptive control with constraints (ACC). Both provide real-time monitoring of process and system parameters. They adjust system and process variables to gain optimum production within specified limits.

Note: The importance of safety in modern machining operations cannot be overemphasized. These operations involve high spindle and cutting speeds, high temperatures, and high cutting forces. Chips produced while machining are hot and sharp and are thrown from the workpiece at high velocities. Although carbide, coated carbide, and cemented carbide inserts are made to withstand the high temperatures and cutting forces of modern machining operations, they will fragment or shatter when subjected to impacts or conditions that exceed their mechanical properties or strengths. Take adequate precautions when working around any machine. Wear the proper safety equipment and clothing. Make sure that all guards and safety equipment are in place and work properly.

Questions and Problems

1. Define the terms cutting speed, spindle speed, and feed rate.
2. Why is it important to program the correct feed rates and spindle speeds for machining operations?
3. Calculate the required spindle speed for a cutting speed of 200 FPM, using a 0.75 in.-diameter milling cutter.
4. Calculate the cutting speed necessary for a spindle speed of 1200 RPM using a 0.25 in.-diameter milling cutter.
5. Calculate the required feed rate in IPM for a spindle speed of 1000 RPM, a chip load of 0.002 chip per tooth, using a four-flute milling cutter.

6. Calculate the feed rate in IPR for a lathe running at 1200 RPM using a 30 IPM feed rate.
7. Calculate the spindle speed and finish feed rate in IPM for a lathe using a 100 FPM cutting speed on a 2 in.-diameter piece of mild steel stock (0.005 IPR finish feed rate).
8. Complete the following table of TLO values.

Tool	Length (in.)	TLO (in.)
1	1.638	
2	2.045	
3	3.122	
4	3.255	
5	3.758	0.5

9. Why are tool length offsets entered or programmed?
10. When would you use tool radius compensation?
11. What is the procedure for using tool radius compensation?
12. Define the term tool nose radius.
13. Why is it important to use TNR compensation?
14. What information is used to determine the TLO for cutting tools on CNC turning machines?
15. What is adaptive control?
16. List the two main categories of adaptive control.
17. What are the advantages of adaptive control systems?
18. List the two general classes of tool changers.
19. List five basic types of automatic tool changers.

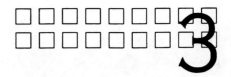

Programming Considerations

Chapter Objectives

After studying this chapter, the student will be able to

- ☐ Describe the use of the Cartesian coordinate system for determining rectangular coordinates.
- ☐ Recognize the common axes of machine movements.
- ☐ Define point-to-point and linear-cut programming.
- ☐ Define continuous path programming.
- ☐ List the steps involved in developing a part program manuscript.
- ☐ Recognize setup sheets for milling and turning machines.
- ☐ Recognize and describe the format details for milling and turning machines.
- ☐ List and define the common word addresses used in CNC programming.
- ☐ Describe absolute programming.
- ☐ Describe incremental or relative programming.
- ☐ Define the use of the zero shift and zero offset.

3.1 Introduction

This chapter will describe the fundamentals of NC/CNC programming on which further programming details and applications are based. Included in this chapter are preliminary programming concepts aimed at developing a strong programming foundation from which to build further concepts and programming experience. These concepts include the Cartesian coordinate system, axes of machine movement, point-to-point, linear-cut, and continuous path or contour programming, the steps involved in developing part programs, and format details.

FIGURE 3.1
CARTESIAN COORDINATE SYSTEM

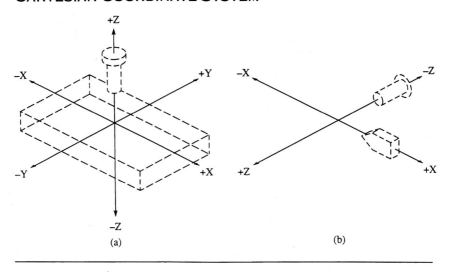

FIGURE 3.2
QUADRANTS AND COORDINATES

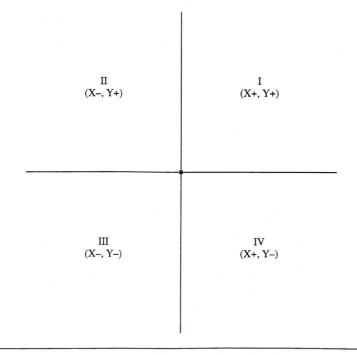

3.2 Cartesian Coordinate System

The **Cartesian coordinate system,** illustrated in Fig. 3.1, is the basis for all machine movement. Within the context of numerical control, an axis is a direction of possible programmed machine movement. For example, on a two-axis milling machine, the X axis serves as the direction of table travel, while the Y axis serves as the direction of saddle travel. On a three-axis mill, the Z axis serves as the direction of spindle travel.

The Cartesian system contains four quadrants: I, II, III, and IV, shown in Fig. 3.2. Notice that the signs (positive and negative) change as you rotate through the quadrants (Fig. 3.2). Each point may be defined by an X, Y, and Z value in the form (X, Y, Z). These values are called **rectangular coordinates** in part programming.

3.3 Axes of Machine Movement

In numerical control, an **axis** is any direction of programmed machine movement based on the machine's slide movements such as the X, Y, and Z axes. The number of directions of simultaneous slide movements or simultaneous axis control is used to classify the machine. Therefore, machines are designated as two-, three-, four-, and five-axis machines, although we live in a three-dimensional world. Positive and negative designations are arbitrarily assigned based on spindle movement. Fig. 3.3 shows one way of assigning positive and negative directions to the axes. The table on a vertical mill, for example, would travel in the opposite direction than that shown. This is because spindle movement defines the sign of the motion direction, not table movement. On milling machines,

FIGURE 3.3 ◻◻◻◻◻◻◻◻◻◻◻◻◻◻◻◻◻◻◻◻◻◻◻◻◻◻◻
AXES OF MACHINE MOVEMENT

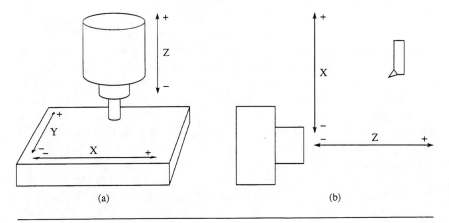

FIGURE 3.4
FIVE-AXIS MACHINE CONFIGURATIONS

The following comprise the fourth and fifth axes in five-axis configurations.

movement toward the workpiece is normally designated as the −Z direction and movement away from the workpiece is +Z. Think of the magnitude of the distance from the tool to the workpiece. As it decreases, the machine travels toward the workpiece—motion in the −Z direction. As the spindle retracts, the distance between the tool and the workpiece increases; the motion is in the +Z direction. On a turning machine, the Z axis generally runs parallel to the chuck face. In this situation, movement in the −Z direction is toward the chuck face, while movement in the +Z direction is away from the chuck face.

The MCU on two-axis NC/CNC machines is capable of controlling simultaneous motion in two axes (generally X and Y for milling machines and X and Z for turning machines). The MCU on three-axis machines is capable of controlling three axes simultaneously (generally X, Y, and Z). Four- and five-axis machines are equipped with MCUs that can control four or five axes simultaneously. Movements not controlled by the MCU, such as adjusting the spindle or knee, are performed manually. There is a two and a half-axis machine, on which the MCU controls the X and Y axes simultaneously, allowing linear control in the Z axis. The fourth axis on milling machines is usually a rotating or tilting machine table, which can rotate or tilt at a controlled rate for additional **contouring** flexibility. Rotating and tilting work tables that are not numerically controlled are available; these index to a number of positions, but are not controlled by the MCU. The fifth axis on milling machines normally describes swivel of the machine head at a controlled rate. This feature manipulates the angle of the cutter, positioning it perpendicular to a point of tangency on a curved or swept surface. Fig. 3.4 illustrates three possible configurations of five-axis milling machines.

There are actually six possible motions, including rotation about the X, Y, and Z axes. The additional motions are designated a, b, and c, respectively. Looking in the positive direction relative to the **origin,** pos-

FIGURE 3.5
SIX POSSIBLE AXES OF MOTION ON MILLING MACHINES

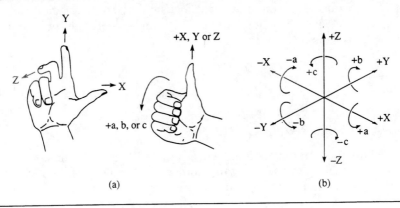

(a)   (b)

itive rotation is in the clockwise direction. Fig. 3.5 illustrates the six possible axes of motion on milling machines.

The four and fifth axes of motion are given here so that you may recognize them. Examples in this text will center on the predominant two and a half- and three-axis milling machines. Four- and five-axis milling machines and machining centers are becoming more common in manufacturing, but two and a half- and three-axis machines are still the most common.

Common varieties of lathes and turning centers include two-, three-, and four-axis machines. The MCUs on two-axis turning machines are capable of controlling movement in the X and Z axes simultaneously. The

FIGURE 3.6
FOUR POSSIBLE AXES OF MOTION ON TURNING MACHINES

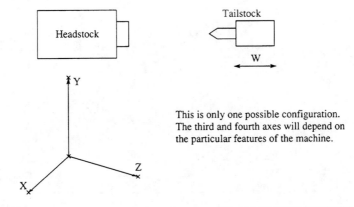

This is only one possible configuration. The third and fourth axes will depend on the particular features of the machine.

third axis on turning machines is a vertical axis often used with milling attachments. The fourth axis is generally reserved for programmable tailstocks. The examples given in this text focus on the more common two-axis turning machines. Fig. 3.6 illustrates the four possible axes of motion on turning machines.

3.4 POINT-TO-POINT AND LINEAR-CUT PROGRAMMING

One method of controlling tool movement is **point-to-point,** which is used to move the tool from one point to another point in operations such as drilling, tapping, and punching. All machines are typically equipped with point-to-point control capabilities. When using point-to-point control, the tool travels to the first point. There it performs the programmed operation, then travels to the next point. On arriving at the second point, it performs the programmed operation and travels to the third programmed point. This continues until the program is complete. The home position is a programmed point used for changing tools or workpieces, where the machine normally locates before and after program execution. Fig. 3.7 illustrates point-to-point control.

FIGURE 3.7 ☐☐☐☐☐☐☐☐☐☐☐☐☐☐☐☐☐☐☐☐☐☐☐☐☐☐☐☐
POINT-TO-POINT CONTROL

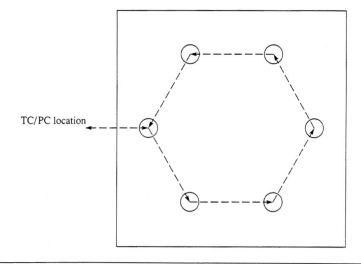

FIGURE 3.8
LINEAR-CUT CONTROL

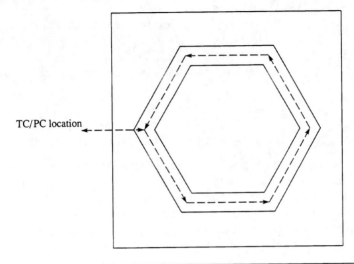

In point-to-point operations, the tool is not in constant contact with the workpiece. It travels from point to point and performs the required operations. After the programmed operation is performed, the spindle retracts to some clearance plane and the table repositions for the next point.

Some machines are capable of only point-to-point control; the MCU for these machines, including punching machines, drilling machines, and some layout devices, is equipped only for point-to-point control. These machines are not capable of linear or circular interpolation. Note that there is a difference between point-to-point controllers and the point-to-point mode. Point-to-point controllers or MCUs are capable only of point-to-point control of the cutting tool. However, point-to-point mode of control can be evoked from an MCU that is capable of other modes of operations.

Another type of tool control is linear-cut, or straight-line. Linear-cut control provides limited contouring abilities; it can perform straight-line milling in either the X or Y axis and straight, angular cuts in the X-Y plane by proportionally controlling the X and Y axes simultaneously. Fig. 3.8 illustrates linear-cut control.

Most machines are capable of linear-cut control. Consider the manual milling machine. It is a simple process to cut a square frame in a piece of stock. But now consider the effort required to cut a diagonal groove from one corner of the square to the opposite corner. It is possible on a manual machine. But remember that one advantage of NC/CNC machines is that they calculate the slope of the line required and proportionately control the axes for you, providing an accurate and reliable cut that can be repeated for each workpiece required.

Point-to-point and linear-cut tool control are both commonly used in NC/CNC machining. However, not all workpieces can be completed using point-to-point and linear-cut control. The programming of arcs, arc segments, and curves requires more sophisticated tool control.

3.5 CONTINUOUS PATH PROGRAMMING

Continuous path tool control systems also have the ability to control two or more axes at the same time. They use this ability to proportionately control axes to cut arcs, arc segments, and complex curves. Generally, MCUs with continuous path control capabilities are also able to perform point-to-point and linear-cut operations. Their primary benefit is the ability to machine more complex features in the workpiece, as shown in Fig. 3.9.

These three types of tool control (point-to-point, linear-cut, and continuous path) provide the necessary flexibility and capability required to machine any workpiece from the simplest to the most complex.

FIGURE 3.9 ◻◻◻◻◻◻◻◻◻◻◻◻◻◻◻◻◻◻◻◻◻◻◻◻◻◻◻◻◻
CONTINUOUS PATH OR CONTOURING CONTROL

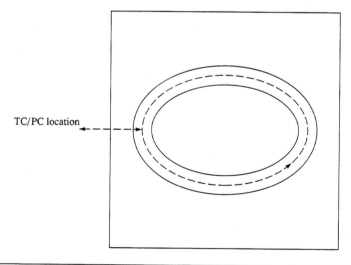

3.6 Developing a Part Program

There are many ways to arrive at a finished program. There is no absolute procedure in developing a part program. Many steps suggested here may be combined or eliminated, as needed. The following generic procedure is a guide, not the rule, in developing CNC programs.

1. Study the blueprint and specifications of the part to be programmed. Work with the engineers to agree on simplifying dimensioning so that programming is easier and more productive.
2. Examine the specifications of the material and the desired finish from the blueprint. These are important features to consider when selecting tooling and programming cutting feeds, speeds, and the number and depth of roughing and finishing cuts.

**FIGURE 3.10A □□□□□□□□□□□□□□□□□□□□□□□□□
EXAMPLES OF SETUP SHEETS**

Mill Setup Instructions

Part Identification								
Oper #	Tool #	Description	Length	TLO	Tool descr.	RPM	Feed	Notes

X zero location
Y zero location
Z zero location

Setup information and sketch

(a)

FIGURE 3.10B ☐☐☐☐☐☐☐☐☐☐☐☐☐☐☐☐☐☐☐☐☐☐☐☐☐☐☐
EXAMPLES OF SETUP SHEETS

Lathe Setup Instructions

Part Identification							
Oper #	Tool #	Offset #	Position	X gage length	Z gage length	Remarks	

X zero location

Z zero location

Setup information and sketch

(b)

---

3. Fill out a setup sheet, showing all work that must be done to finish the part. Include a sketch of the tool change position, part change position, and other necessary information. Keep in mind that other people may need to work from the information you provide; think of what you would want to know if you were in their place. Fig. 3.10 shows an example of a setup sheet for a milling and turning machine.

4. Use the previous information to rough out a program on a form called a **manuscript**. After completing a rough draft, go over the manuscript again to determine whether you can edit it, in terms of program steps and execution time. You may want to go over it more than once; you can often trim a program down each time you go through it. You cannot trim programmed steps and execution time by exceeding recommended feed rates and speeds or practicing unsafe procedures. When programming, do the job correctly and safely.

FIGURE 3.11 ◻◻◻◻◻◻◻◻◻◻◻◻◻◻◻◻◻◻◻◻◻◻◻◻◻◻◻
FLOWCHART OF BASIC PROGRAMMING PROCEDURES

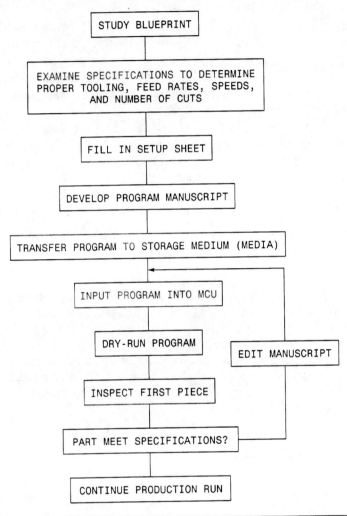

---

5. Type the program into the computer or keypunch the program on tape. Check to make sure the program is correct. *Now* is the time to discover a typo.
6. Load the program into the MCU.
7. Dry-run the program. This is normally performed with the table lowered safely away from the spindle and with the spindle turned off. The purpose of the dry run is to catch any errors that may affect the proper execution of the program.

8. Inspect the first piece produced by the machine to determine if the part meets specifications on dimensions and finish. If the piece is acceptable, continue to run the part.
9. Check to make sure the speeds and feeds are correct as recommended. Make any necessary corrections and rerun the part.
10. Check periodically to ensure that the part produced is still acceptable. A CNC machine is an accurate machine, but relies on the operator for proper tooling and setup. Fig. 3.11 is a flow chart of the basic programming procedures.

Because the success of the production run hinges on the manuscript, the programmer can take satisfaction from completing a successful manuscript. However, if problems occur, it is often due to errors in the programmed information. Take a minute and study the flow chart in Fig. 3.11. Acquaint yourself with the basic procedures involved in programming. Once you feel comfortable with these steps, we can build on and study these steps in greater detail.

3.7 FORMAT DETAIL

There are many different types of MCUs currently available, and each different type may use a different format detail. A format detail describes the appearance and arrangement of values the MCU will accept. The programmer should be acquainted with the format detail of the MCU before beginning programming. The following list of code words explains the more commonly used words and their meanings. Also discussed is the order and format of the programmed words. Two format detail examples are given, one for a lathe and one for a mill.

TYPICAL FORMAT DETAIL FOR A LATHE

```
N4 G2 X+3.4 Z+3.4 I+3.4 K+3.4 P3.3 Q3.3 F3.1 S4 T2 M2
      U+3.4 W+3.4                            F0.4
```

where the words

N (sequence number): allows four unsigned digits.

G (preparatory functions): allows two unsigned digits.

X and Z (primary motion dimension words): accepts signed numbers having three digits to the left and four digits to the right of the decimal point.

U and W (secondary motion dimension words): accepts signed numbers having three digits to the left and four digits to the right of the decimal point.

I and K (auxiliary words): same format as X and Z.

P and Q (dwell times and canned cycles): allows three digits to the left and right of the decimal.

F (feed rates): IPM programming allows two digits to the left and right of the decimal, while IPR programming allows only four digits to the right.

S (speed functions): allows four unsigned digits.

T (tool functions): four digits: the left pair is the tool number, the right pair is the tool offset register number. For example, T0608 specifies tool #6 and offset register #8.

M (miscellaneous functions): allows two unsigned digits.

TYPICAL FORMAT DETAIL FOR THE MILLING MACHINE

N5 G2 X+3.4 Y+3.4 Z+3.4 I+3.4 J+3.4 K+3.4 F3.1 S4 T2 M2
                                                F0.4

where the words

N (sequence number): allows four unsigned digits.

G (preparatory functions): allows two unsigned digits.

X, Y and Z (primary motion dimension words): accepts signed numbers having three digits to the left and four digits to the right of the decimal point.

I, J, and K (secondary motion dimension words): accepts signed numbers having three digits to the left and four digits to the right of the decimal point.

F (feed rates): IPM programming allows two digits to the left and right of the decimal, while IPR programming allows only four digits to the right.

S (speed functions): allows four unsigned digits.

T (tool functions): four digits: the left pair is the tool number, the right pair is the tool offset register number. For example, T0608 specifies that tool #6 should be called and offset register #8 holds the necessary offset.

M (miscellaneous functions): allows two unsigned digits.

The following table briefly summarizes common words and uses of these words.

Word	Use
N4	Sequence number. Informational rather than functional. Each block of information in a program begins with a block sequence number (N). Block sequence numbers add readability to the program and are helpful in editing.

Word	Use
G2	G codes. Used for machine control. G codes are used to initiate motion and describe the motion procedure. For instance, G1 and G0 are used to initiate movement at programmed feed rates and rapid rates, respectively. Two other examples are G90 and G91, which are used to inform the MCU that absolute and incremental positioning will be used, respectively. Other G codes initiate the use of canned cycles, zero shift, dwell periods, and other features relating to machine control. Descriptions of the various G codes and their functions are found in Appendix A.
	Most G codes are **modal commands.** Modality means that these codes remain in effect until canceled by another code. For example, G90 initiates absolute positioning and is modal. G90 will remain in effect until canceled by a G91 (incremental positioning), for instance. As another example, G1 specifies that the tool should move at the programmed feed rate. This is a modal command that remains in effect until canceled by a code such as the G0 (rapid movement). Modal commands will be specified as such as they are discussed. When applicable, the code choices that cancel these modal codes will also be given.
	Not all commands are modal. Nonmodal commands such as dwell (G4) remain in effect for one block only. These will also be specified in the text, at their point of application and in the appendices. More complete information can be found in the manual that accompanies the MCU. The manual is always the definitive source for programming.
X+3.4	X axis coordinate value. Signed number that specifies the amount and direction of travel in the X axis. The words X, Y, I, J, K, U, and W do not initiate any movement, but contain the intended motion direction and magnitude of that motion. For example, G01X−2.5 indicates that the MCU should move the table 2.5 units in the −X direction at the programmed feed rate. Specifically, these words describe the designated axis values for their intended motion vectors.
Y+3.4	Y axis coordinate value. Signed number that specifies the amount and direction of travel in the Y axis.
Z+3.4	Z axis coordinate value. Signed number that specifies the amount and direction of travel in the Z axis.
R3.4	Used to program the radius of an arc in the radius method of circular interpolation (discussed in later chapters).
I+3.4	Used to program the center of the arc in circular interpolation using the centerpoint method.
J+3.4	Used to program the center of the arc in circular interpolation using the centerpoint method.
K+3.4	Used to program the center of the arc in circular interpolation using the centerpoint method.
A/B/C+3.4	Used to program fourth, fifth, and sixth axis motion. Signed numbers that specify the amount and direction of travel in the appropriate axis. See Fig. 3.4 for an illustration of these axes and their motion directions.

Word	Use
P+3.4	X axis projection of the tool diameter compensation vector. The P word is used to specify the amount of compensation in the X axis. See Chapter 2 for a more complete description of tool diameter compensation. The P word address is also used in canned cycles (discussed in later chapters).
Q+3.4	Y axis projection of the tool diameter compensation vector. Q is used to specify the amount of compensation in the Y axis and in canned cycles.
F0.4 or F3.1	Programmed feed rate for IPR. Programmed feed rate for IPM.
S4	Programmed spindle or chuck speed.
D4	Used to program machine delay times (dwell) in seconds.
T2 or T4	Tool number. T4 (lathe) with offset number accompanying tool number.
M2	Miscellaneous functions. Typically, programs machine functions not related to machine movement, i.e., tool changes, coolant, and end of program. The M00 function stops program execution. M02 is used to designate the end of a program and initiate tape rewind (return to the beginning of the program). M06 is used to initiate tool changes. This also stops the tool

FIGURE 3.12 ◻◻◻◻◻◻◻◻◻◻◻◻◻◻◻◻◻◻◻◻◻◻◻◻◻◻
SAMPLE PART PROGRAM

```
%
/N5G28U0W0
/N10G0U-20000W-10000
N15G92X71722Z53935S2500
N20G90G95G96F80S350T101M13
N25G0X16370Z11660
N30G4P2000
N35G1X-300
N40G0X14370Z12660
N45G4P1000
N50G71P45Q105U200W100D500
N55G0X0
N60G1Z11560
N65G3X13120Z5000I0K-6560
N70G1X14370
N75G0X71722Z53935T100
N80G90G95G96F50S400T202M13
N85G0X13120Z11560
N90G4P2000
N95G1X-500
N100X0
N105G4P1000
N110G3X13120Z5000I0K-6560
N115G1X13500
N120G0X71722Z53935T200
N125M30
%
```

Word	Use
	spindle or chuck. After a tool change, M03 and M04 are used to restart the spindle in the clockwise or counterclockwise direction, respectively. Miscellaneous functions and their purposes are given in Appendix A.

Take a minute to quiz yourself on the meanings of N, G, X, Y, Z, I, J, K, U, W, P, D, F, S, T, and M. Become familiar with the words and format detail of the machines you will be using. These words and formats will be used throughout this text in the examples.

The sample program given in Fig. 3.12 illustrates the use of some of these words in CNC programming. Study the example to reinforce your

**FIGURE 3.13A
ABSOLUTE POSITIONING**

```
%
N100G0G90X-4.Y0F18.0S1800T1M6
N110X2.Y2.Z.1
N120G1Z-.5
N130G0Z.1
N140X6.
N150G1Z-.5
N160G0Z.1
N170Y4.
N180G1Z-.5
N190G0Z.1
N200X2.
N210G1Z-.5
N220G0Z.1
N230X-4.Y0M2
%
```

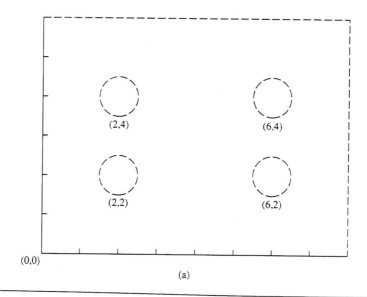

(a)

understanding of the use of these words. Refer to Appendix A for a description of words or codes you are not familiar with at this point.

When an NC/CNC machine is first turned on, it has a set of default codes that it uses until programmed otherwise. The MCU reads the G and M values that have been programmed into its ROM. It establishes these as the default values at startup. When programming, it is important that you be aware of the default values of the MCU. It is also a good programming practice to develop a line of default values that are included as the first line in your programs. Startup or initialization procedures are given with programming examples.

3.8 Absolute Programming

There are two basic methods of positioning the machine with respect to the coordinate system shown in Fig. 3.3: absolute and incremental, or relative, positioning. This section will discuss and provide examples of both methods. Study these examples and compare the two methods. Later, we will discuss the advantages and disadvantages of each.

Figure 3.13B ▫▫▫▫▫▫▫▫▫▫▫▫▫▫▫▫▫▫▫▫▫▫▫▫▫▫▫▫
Absolute Positioning

```
%
N100G0G90X3.Z5.F18.0S1800T0101
N110G1X0Z4.
N120G1X1.
N130Z2.
N140X2.
N150Z0
N160G0X3.Z5.M30
%
```

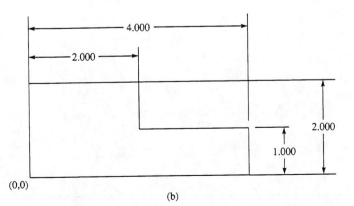

(b)

72　CHAPTER 3　PROGRAMMING CONSIDERATIONS

In absolute positioning, programmed coordinates reference a fixed **zero reference point**. The MCU uses this fixed zero point to position the machine during program execution. An example of absolute positioning is given in Fig. 3.13. In CNC programming, it is not necessary to repeat coordinates that remain unchanged. In addition, positive numbers typically do not require the positive (+) sign. Most machines assume a number to be positive unless designated otherwise by the negative (−) sign.

**FIGURE 3.14A
INCREMENTAL POSITIONING**

```
%
N100G0G90X-4.Y0Z.1F18.0S1800T1M6
N110G91X6.Y2.
N120G1Z-.6
N130G0Z.6
N140X4.
N150G1Z-.6
N160G0Z.6
N170Y2.
N180G1Z-.6
N190G0Z.6
N200X-4.
N210G1Z-.6
N220G0Z.6
N230X-6.Y-4.M2
%
```

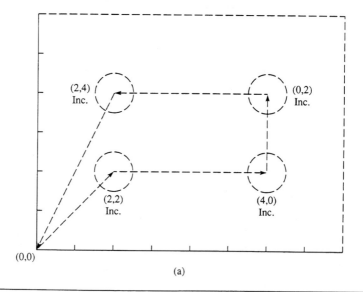

(a)

3.9 Incremental (Relative) Programming

In **incremental**, or **relative, positioning**, programmed movements are based on the change in position between successive points. Each subsequent position is based on the previous position. The relative distance between where the tool is currently located and the next programmed point is entered into the program. This may simplify programming, allowing the programmer to specify the amount of movement required rather than calculating absolute coordinates. The savings in time and effort of the programmer depends on how the feature locations are designated on the part drawing. After each programmed movement, the current position becomes the starting position for the next programmed movement. Positive and negative values are programmed to initiate the desired direction of machine movement. The MCU does not reference a fixed zero point.

Incremental and absolute positioning can be used on both point-to-point and continuous path control systems. Fig. 3.14 provides an example of incremental positioning. (This is the same part shown in Fig. 3.13.)

Compare the two positioning examples (Fig. 3.13 and Fig. 3.14). Can you determine the advantages and disadvantages of each method?

One advantage of absolute positioning over incremental positioning is that positioning errors are compounded in incremental positioning. If a

FIGURE 3.14B ◻◻◻◻◻◻◻◻◻◻◻◻◻◻◻◻◻◻◻◻◻◻◻◻◻◻
INCREMENTAL POSITIONING

```
%
N100G0G90X3.Z5.F18.0S1800T0101
N110G1G91X-3.Z-1.
N120X1.
N130Z-2.
N140X1.
N150Z-2.
N160G0X1.Z5.M30
%
```

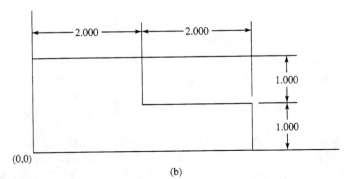
(b)

FIGURE 3.15A
ABSOLUTE AND INCREMENTAL POSITIONING

```
%
N100G0G90X-4.Y0F18.0S1800T1M6
N110X2.Y2.Z.1
N120G1Z-.5
N130G0Z.1
N140G91X4.
N150G1G90Z-.5
N160G0Z.1
N170G91Y2.
N180G1G90Z-.5
N190G0Z.1
N200G91X-4.
N210G1G90Z-.5
N220G0Z.1
N230X-4.Y0M2
%
```

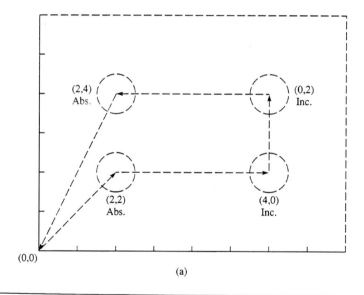

(a)

positioning error occurs in incremental positioning, all subsequent points contain the accumulated error. In absolute positioning, only the incorrect point will be in error since points are programmed from a fixed reference. This is not to say that absolute positioning is better than incremental. These two methods offer flexibility in CNC programming. One method is not better than the other. And both rely on the ability of the programmer.

Both methods can be used in the same program. Fig. 3.15 illustrates the use of both absolute and incremental positioning in the same program.

**FIGURE 3.15B
ABSOLUTE AND INCREMENTAL POSITIONING**

```
%
N100G0G90X3.Z5.F18.0S1800T0101
N110G1X0Z4.
N120G91X1.
N130Z-2.
N140G90X2.
N150Z0.
N160G0X3.Z5.M30
%
```

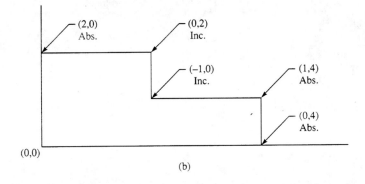

(b)

3.10 Zero Shift

The **zero shift** option is available on most NC machines. Zero shifting allows the relocation of the zero reference of the workpiece to another valid location within the travel limits of the machine. Machine options concerning zero shifting range from no zero shift (using a fixed zero reference) to full range offset to full floating zero capacities.

Fixed zero machines have a permanent zero location that does not allow any change in the zero position. The operator receives instructions on how to set up the workpiece in order for the program to run properly, based on the fixed zero reference. This may be accomplished by locating off a corner or previously cut edge, or by positioning off a known point. Once the workpiece is set up properly, it is firmly clamped to the table and the program is reset and allowed to run to completion. This procedure is required each time a new piece is to be cut. Normally, the workpiece is positioned so that all programmed movements are made in the first quadrant. This keeps all numbers positive in absolute programming. Although you may run across them, fixed zero machines are obsolete.

Full-range zero offset may be available on fixed zero machines. The controller retains the permanent zero location, while allowing another

FIGURE 3.16A □□□□□□□□□□□□□□□□□□□□□□□□□□
Resetting Absolute Zero

```
%
N100G0G90X-4.Y0F18.0S1800T1M6
N110X4.Y3.
N120G92X0Y0    If the cutter is located over the new zero
|              location.
|
|
|
|
%
```

If the cutter is not located over the new zero location, the X and Y values from the new zero to the current tool position are entered.

```
%
N100G0G90X-4.Y0F18.0S1800T1M6
N110X0Y0
N120G92X-4.Y-3.
|
|
|
|
%
```

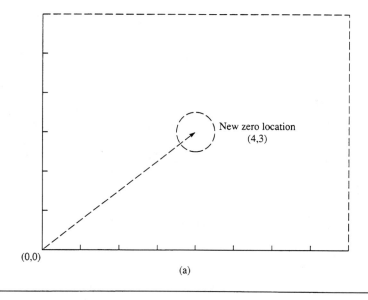

(a)

zero location to be used during program execution. The machine zero location is still located in a permanent position, but the full-range offset capability allows locating the workpiece at any convenient location on the table within travel limits of the machine axes.

FIGURE 3.16B
RESETTING ABSOLUTE ZERO

```
%
N100G0G90X3.Z5.F18.0S1800T0101
N110G92X3.Z1.      Distance from the new zero to the current
|                  tool location.
|
|
|
%
```

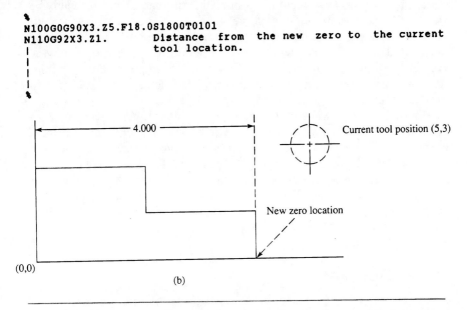

(b)

The workpiece is clamped to the table. The table is then jogged (moving the table using the jog feature of the machine) over to the new zero location and the machine axes rezeroed or new position coordinates entered into the MCU. The controller remembers the fixed zero reference location while using the new coordinates throughout program execution. The machine will remember the new coordinates until programmed otherwise or until machine power is removed. Most MCUs have battery backup for their memories, which will retain values for a limited period after power removal. Fixed zero offset machines are also obsolete.

Full floating zero machines have no fixed reference location. A zero reference point must be established for each new setup. The workpiece may be firmly clamped to the table at any convenient position on the table within the travel limits of machine axes. The new setup coordinates are then entered into the controller. The programmer usually gives these setup coordinates to the operator on the setup sheet. This allows the programmer to select any convenient zero reference point while programming. Programmed coordinates reference the programmer's zero reference rather than any machine-dependent coordinates. This allows the programmer to use all four quadrants.

The machine is positioned so that the tool or spindle is located over the setup point. The first two instructions in the program usually position the machine and reset the zero point. The readout should then read the new programmed coordinates, i.e., (0,0). These new coordinates provide a reference for all subsequent programmed coordinates. The zero may be shifted later in the program using the same procedure. Full floating zero operations reduce the required setup time. The zero reference point may be located at any convenient point throughout the program, allowing greater flexibility in programming. However, subsequent setups require rezeroing the machine, if the workpiece cannot be accurately located in each new setup.

During execution of a program, the G92 code is used to reset the zero reference for absolute programming. The programmer may elect to program the G92 code followed by the distance from the new zero location to the current tool position. Fig. 3.16 illustrates the use of the G92 code. Zero shifting allows greater program flexibility. It may be used so that programmed values are all positive in absolute programming or, in the case of mirroring (discussed later), where you need a central reference to reflect across.

3.11 SUMMARY

The three types of NC control systems are point-to-point, linear-cut, and continuous path or contouring. Machine movement references the Cartesian coordinate system. Any point within the Cartesian coordinate system can be located with an (X,Y,Z) coordinate. There are six possible machine motions, and each machine motion corresponds to a machine axis. Machines capable of simultaneous control in three axes are called three-axis CNC machines. Machines capable of two-axis simultaneous control and linear control in a third axis are known as two and a half-axis machines. Machines capable of simultaneous control in more than three axes are designated four-axis and five-axis CNC machines. The format detail of an MCU designates the available words and value formats for that MCU. Absolute positioning references a fixed zero point, while incremental positioning references machine movements relative to previously programmed points. Zero shift repositions the zero reference to a convenient location on the workpiece.

QUESTIONS AND PROBLEMS

1. What are axes of machine movement?
2. Draw a sketch of the three primary axes of machine movement on the mill and the two primary axes on the lathe.
3. Describe the differences between the point-to-point, linear-cut, and continuous path control systems.

FIGURE 3.17 ◻◻◻◻◻◻◻◻◻◻◻◻◻◻◻◻◻◻◻◻◻◻◻◻◻◻◻◻
PROBLEM #8

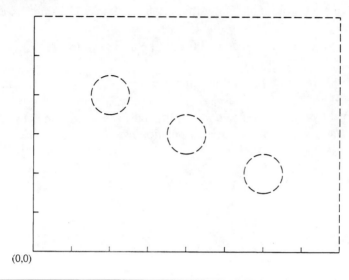

4. Describe the purpose of the format detail for an MCU.
5. Define the following words: N, G, X, Y, Z, I, J, K, U, W, P, D, F, S, T, and M.
6. What are the two types of positioning systems and how are they different?
7. Explain the use of the zero shift option.
8. Write a program to drill the three holes pictured in Fig. 3.17. Use absolute positioning.

FIGURE 3.18 ◻◻◻◻◻◻◻◻◻◻◻◻◻◻◻◻◻◻◻◻◻◻◻◻◻◻◻◻
PROBLEM #11

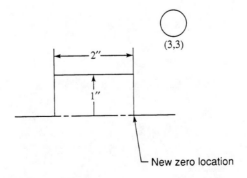

9. Write a program to drill the same three holes in Fig. 3.17 using incremental positioning.
10. Write a program to drill the three holes in Fig. 3.17 using absolute and incremental positioning.
11. Write a program to face and turn the part shown in Fig. 3.18 using absolute positioning.
12. Write a program to face and turn the part shown in Fig. 3.18 using incremental positioning.
13. Write a program to face and turn the part shown in Fig. 3.18 using both absolute and incremental positioning.

Input Media, Formats, and Program Transfer

Chapter Objectives

After studying this chapter, the student will be able to
- List and describe the various forms of input media.
- Recognize and describe the RS–244 and RS–358 codes.
- Describe the different ways of entering numbers into the MCU.
- List and describe the various programming formats.
- Describe the different methods and equipment used in program transfer.

4.1 Introduction

This chapter will focus on the various media, formatting techniques, and transfer methods used to input programs into numerical control machines. These programming media store the required information in a semipermanent form until needed. When required, the media is retrieved and fed into the controller. Modern numerically controlled machines offer a variety of ways to input media including manual data input (MDI), punched tape, magnetic media, and direct interfacing.

4.2 Input Media

Machines programmed manually use the keypad found on most controllers. This keypad resembles a calculator keypad and contains the necessary characters and symbols to write programs for the MCU. Fig. 4.1 shows a typical MDI keypad. This text will center on the word address programming format. Briefly, the word address format uses a unique letter address for each value. During MDI programming, the word address

Figure 4.1
MDI Keypad

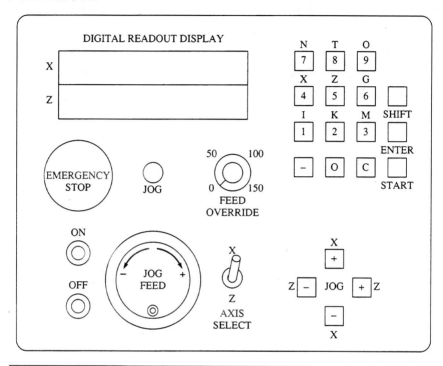

is entered, followed by the value for that address. Values of zero or that remain unchanged are not repeated. Depending on the MCU capabilities, prompts giving the word address appear, and values are entered corresponding to the word address. Most machines allow for downloading an MDI-produced program to suitable storage through the RS-232 interface (discussed later in this chapter).

The MDI method of programming eliminates the need for an intermediate step in programming and allows for program editing directly on the machine. While MDI can be slow and tedious, background editing features allow the programmer to edit or create a part program while another part program is running, saving some time.

NC and CNC machines also receive programming information from 1-in.-wide tape perforated with holes. Common tape materials include paper, oil-resistant paper, Mylar-paper laminations, aluminum, and other suitable materials. Most of these materials are available in a wide variety of colors, depending on individual preference or need. The color of the tape also affects the ability of light to pass through it. If the tape reader uses a strong light source, the light may pass through the tape as well as the holes in the tape, causing the reader to misread the coded values. In such cases, darker tape materials are used to prevent the light from passing through the unperforated tape.

Paper tape as an input medium has advantages and disadvantages. One advantage of paper is its low initial cost. With special treatments to make it oil- and water-resistant, paper tape is a short-term cost-efficient coding material. However, it proves to be the most expensive over the life of the machine tool. The primary disadvantage of paper tape as an input medium is that it is less durable than other materials, tearing and staining easily. Manufacturers prefer aluminum or Mylar laminates in heavier production schedules. They are more expensive, but more durable than paper. Another disadvantage common to all punched tape use is its lack of reusability. A punched tape cannot be altered. Paper is an expendable commodity and less reliable than magnetic media, especially floppy disks. For these reasons, paper tape use has declined in favor of floppy diskettes. Paper tape is normally reserved for older machines without diskette capability or as a backup method.

Standard tape materials are available in inch-wide rolls that contain between 1000 and 2000 ft. (300 to 600 m). The width of the tape is 1.000 +/− 0.003 in. (25.4 +/− 0.076 mm) and typical thicknesses range from 0.004 +/− 0.0003 in. (0.10 +/− 0.008 mm). The inch-wide tape allows for eight columns or tracks of holes per row plus a column of holes produced by the feed sprockets within the punch. The feed sprocket in the reader uses the sprocket holes in the tape to feed the tape over the photoelectric equipment. Fig. 4.2 provides tape specifications.

Notice that the tape in Fig. 4.2 has a series of holes oriented in rows and columns. The tape punch/reader contains a row of eight punches enclosed over a channel through which the tape feeds. The tape punch responds to external input and activates the required punches in response to the input. The external input may come from a number of sources, including a downloaded file from the computer, keyed input from a keypunch machine, or an MDI file downloaded from the MCU.

Alternatives to dedicated punched tape are magnetic tapes and floppy disks. Magnetic tapes are usually polymeric materials coated with iron oxide and resemble standard audio recording tapes. They are most often quarter-inch computer-grade cassette tapes. Magnetic tapes offer good data protection and a convenient size for handling and storage. Instead of punching holes in these tapes, the computer magnetizes the information onto the tape. When the magnetic tape passes over a transducer in the controller's tape reader, the signal from the transducer reproduces the coded information on the tape. The signal from the transducer is very small and requires amplification before the MCU can interpret it properly. Transducers convert the magnetic signals on the tape, containing the binary form of the source program, into electrical pulses for amplification.

FIGURE 4.2
TAPE SPECIFICATIONS

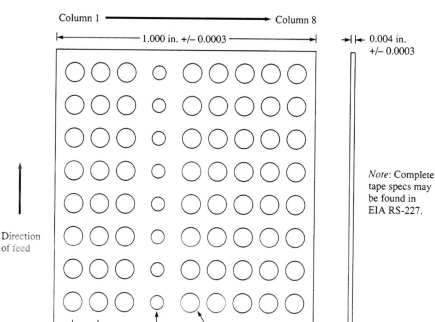

Transducers simply convert data from one form to another and do not interpret the data.

Magnetic tapes store much more data per unit of length than paper tapes. They also transfer data faster and more reliably than paper tapes. However (there is no such thing as a perfect input medium), magnetic tapes require careful handling. Strong electric or magnetic fields passing over them erase the data contained on the tapes. Their reusability provides an advantage, but if a tape is accidentally erased, the data contained on it is lost. Magnetic tapes are also easily damaged by oil, dirt, and grease in the environment. Fingerprints or foreign matter on the tape may alter the information it contains, rendering it useless. Magnetic tapes are commonly used for downloading programs from off-line terminals.

The input medium of choice is the floppy diskette. Due to the widespread availability and use of computers, floppy disks have largely replaced other methods of data storage. Floppy disks have the same handling precautions as magnetic tapes; they are susceptible to dirt, grease, and other foreign materials. They have the advantage of providing the most cost-effective storage of part programs. Floppy disks are reusable, having the capability to be altered repeatedly without a significant loss in data integrity. Computer files (programs containing usable data) are deleted and new part programs stored on the disk. Floppy disks are transferrable from one computer to another of the same type, e.g., IBM and compatibles, Macintosh, and other computer types. Based on their low cost, reliability, availability, portability, and reusability, floppy disks are the most prevalent method of storing and transferring part programs in the CNC environment.

Who knows what the future holds for CNC input media? Consider the advances in data storage made within the last few years. Will the use of compact discs increase in CNC applications? Recent technology advances in CD-ROM provide optical storage capabilities that are erasable. Compact discs are capable of storing millions of pieces of information; they also provide increased data protection. What about optical readers, such as UPC code readers, that code and interpret data? Can these find an application in CNC operations? Voice recognition systems are a contemporary technology that also provides possibilities in CNC technology. Consider the recent work done in neural networks. Every day computers come closer to being able to think for themselves.

The following list provides the advantages and disadvantages of each of the input media previously discussed.

1. MDI allows fast editing of previously input programs, yet ties up the machine.
2. Paper has the lowest initial cost, yet proves to be the highest long-term investment. In addition, paper tapes are the least reliable input medium available.
3. Magnetic tape is capable of storing vast amounts of data, but is susceptible to strong magnetic and electric fields, dirt, grease, and other foreign matter. In addition, magnetic tapes are erasable and reusable.
4. Floppy disks are the medium of choice. They provide low-cost, reliable, and fast data access and storage. While they are also susceptible to strong magnetic and electric fields, rough handling, and harsh environments, with proper care floppy disks provide the cheapest, most efficient input medium.

4.3 CODING SYSTEMS—RS–244 AND RS–358

In order for the MCU to interpret programmed information, some type of **coding system** is necessary. The programmer inputs information into a source program. The source program is then coded into input that the MCU can read. The MCU accepts information in binary form. A programmer can program the information in binary form or use available coding equipment to speed this process considerably.

One piece of coding equipment is the computer. Information is fed into the MCU as **bits** and **bytes**. Computers and computer-controlled equipment work with binary digits, not the letters and numbers we are more familiar with. The binary system contains only two states. The MCU uses a series of on and off (or high and low or a 1 and a 0 states) conditions to control machine functions. In punched tapes, the holes represent the "on" or "1" condition and the absence of a hole represents the "off" or "0" condition. A presence of magnetized particles on magnetic media represents a 1, while the absence represents a 0. Each row on the tape has space for eight holes or bits. Those familiar with computer technology will recognize there are eight bits in a byte. Each row on the tape corresponds to one byte of information.

To code the information properly, a coding system is applied to the source program. For example, the programmer may sit down at the computer terminal and create the source program using a text editor, word processor, computer-aided manufacturing (CAM) program, or other computer program. The computer stores the information in binary form (bits and bytes) onto the proper medium, such as a floppy disk. The computer is then capable of copying the file through a serial interface (discussed in this chapter in the section on program transfer) to a tape punch or MCU. The tape punch responds to the input from the computer by activating the correct punches. These punches perforate the tape to represent each byte of information sent from the computer. A program transferred to the MCU goes directly into the controller's memory.

Coding systems make programming easier. The computer interprets the information written by the programmer into binary form, simplifying the coding process considerably. Another important point presented in the previous example is that the computer and the tape punch or MCU must speak the same language (use the same coding system) for them to effectively communicate.

As an introduction to coding systems, consider the **binary,** or base-2, **numbering system.** This system uses the number 2 raised to a power. Looking at Table 4.1, starting on the right and traveling to the left, each column increases the exponent by a factor of 1.

A simple, frequently used coding system is the **binary coded decimal** or **BCD** system. Electronic Industries Association (**EIA**) standard RS–244 covers this coding system. The EIA has assigned a particular configuration to each character, number, and symbol used in coding information based on the BCD system. The BCD system uses the first four columns of the binary system. The RS–244 coding system is capable of representing the numbers from 0–15 (0000 to 1111). However, part features and drawings are based on the decimal system, which uses the digits 0–9.

SECTION 4.3 CODING SYSTEMS—RS–244 AND RS–358

TABLE 4.1
BINARY NUMBER SYSTEM

Column or Track	8	7	6	5	4	3	2	1
Power of 2	7	6	5	4	3	2	1	0
Value of Number	128	64	32	16	8	4	2	1
Example: 243	1	1	1	1	0	0	1	1

```
243 = 128
       64
       32
       16
        2
      + 1
```

Therefore, the numbers 10–15 are not used to code program information. Table 4.2 gives an example of BCD coding.

Another commonly used coding system is a subset of the American Standard Code for Information Interchange, or **ASCII,** code. The ASCII code was created by the American National Standards Institute (ANSI) to provide an international standard for information processing and communications. EIA standard RS–358 refers to this subset of the ASCII code. The RS–244 (EIA) code differs from the RS–358 (ASCII) code. Table 4.3 compares the two codes. Study the two coding systems and try to determine some of their differences. Most machines today can use either code.

Track 5, or column 5, on an NC/CNC tape controls the **parity** of the tape. Parity is an equality of condition between subsequent values. In other words, the same condition or state exists for all values. In part programming, parity checking determines if each byte contains an even or odd number of bits. It is important to check for parity to help eliminate errors or mistyped information.

Parity is checked by a parity bit in the fifth column. This column maintains an even or odd number of holes in each row (in even or odd parity checking, respectively). For instance, coding the number 6 requires only two holes, 0000 0110. Without the parity bit, we would have an even number of holes in the row. Using odd parity checking, the machine would interpret 0110 as an error and stop reading at that point. The reader would interpret the even number of holes as an error in the tape and return an error message. Therefore, the fifth track is used to add a hole in the row to come up with an odd parity, i.e., 0001 0110. The RS–244 coding method uses odd parity checking, while RS–358 uses an even parity check. The RS–244 system uses the parity bit to maintain an odd number

TABLE 4.2 BINARY CODED DECIMAL SYSTEM

Character	EOL 8	X 7	0 6	P 5	8 4	4 3	2 2	1 1 Track
0				0				
1								0
2							0	
3				0			0	0
4					0			
5				0		0		0
6				0		0	0	
7						0	0	0
8					0			
9				0	0			0
a		0	0					0
b		0	0				0	
c		0	0	0			0	0
d		0	0			0		
e		0	0	0		0		0
f		0	0	0		0	0	
g		0	0			0	0	0
h		0	0		0			
i		0	0	0	0			0
j		0		0				0
k		0		0			0	
l		0					0	0
m		0		0		0		
n		0				0		0
o		0				0	0	
p		0		0		0	0	0
q		0		0	0			
r		0			0			0
s			0	0			0	
t			0				0	0
u			0	0		0		
v			0			0		0
w			0			0	0	
x			0	0		0	0	0
y			0	0	0			
z			0		0			0
. (period)		0	0		0		0	0
, (comma)			0	0	0		0	0
/ (slash)			0	0				0
+ (plus)		0	0	0				
− (minus)		0						
space				0				
delete		0	0	0	0	0	0	0
EOB (CR)	0							
backspace			0		0		0	
tab			0	0	0	0	0	
EOR				0			0	0
uppercase		0	0	0	0	0		
lowercase		0	0	0	0		0	

TABLE 4.3   RS–244 AND RS–358 CODES

Character	| RS–244 (EIA) |||||||||									| RS–358 (ASCII) |||||||||								
	8	7	6	5	4	.	3	2	1	8	7	6	5	4	.	3	2	1
0			0			·						0	0		·			
1						·			0			0	0		·			0
2						·		0				0	0		·		0	
3				0		·		0	0			0	0		·		0	0
4						·	0					0	0		·	0		
5				0		·	0		0			0	0		·	0		0
6				0		·	0	0				0	0		·	0	0	
7						·	0	0	0			0	0		·	0	0	0
8					0	·						0	0	0	·			
9				0	0	·			0			0	0	0	·			0
a		0	0			·			0		0	0			·			0
b		0	0			·		0			0	0			·		0	
c		0	0	0		·		0	0		0	0			·		0	0
d		0	0			·	0				0	0			·	0		
e		0	0	0		·	0		0		0	0			·	0		0
f		0	0	0		·	0	0			0	0			·	0	0	
g		0	0			·	0	0	0		0	0			·	0	0	0
h		0	0		0	·					0	0		0	·			
i		0	0	0	0	·			0		0	0		0	·			0
j		0		0		·			0		0	0		0	·		0	
k		0		0		·		0			0	0		0	·		0	0
l		0				·		0	0		0	0		0	·	0		
m		0		0		·	0				0	0		0	·	0		0
n		0				·	0		0		0	0		0	·	0	0	
o		0				·	0	0			0	0		0	·	0	0	0
p		0		0		·	0	0	0		0	0	0		·			
q		0		0	0	·					0	0	0		·			0
r		0			0	·			0		0	0	0		·		0	
s			0	0		·		0			0	0	0		·		0	0
t			0			·		0	0		0	0	0		·	0		
u			0	0		·	0				0	0	0		·	0		0
v			0			·	0		0		0	0	0		·	0	0	
w			0			·	0	0			0	0	0		·	0	0	0
x			0	0		·	0	0	0		0	0	0	0	·			
y			0	0	0	·					0	0	0	0	·			0
z			0		0	·			0		0	0	0	0	·		0	
. (period)		0	0		0	·		0	0			0		0	·	0	0	
, (comma)			0	0	0	·		0	0		0			0	·	0	0	
/ (slash)			0	0		·			0		0			0	·	0	0	0
+ (plus)		0	0	0		·						0		0	·		0	0
− (minus)		0				·						0		0	·	0		0
space				0		·					0	0			·			
delete		0	0	0	0	·	0	0	0		0	0	0	0	·	0	0	0
EOB	0					·					0			0	·	0		0
backspace			0		0	·		0			0			0	·			
tab			0	0	0	·	0	0						0	·			0
EOR				0		·		0	0				0	0	·			0

of bits in each byte. The RS-358 coding method uses the parity bit to maintain an even number of bits for each byte.

Take a few minutes to try to code the statement N0100 X1 Y5 Z0 using the RS-244 and RS-358 coding systems given in Table 4.3. Remember to include parity checking.

Not all coded information is an alphanumeric character or symbol; the tab code is one exception. The tab code is used to identify or arrange information in a desired order. A tab is just a nonprinting spacing requirement used to separate characters. A tab code is easily recognizable on the tape by a series of holes in tracks 2, 3, 4, 5, and 6. Tab codes can make the program more readable by making tabulated columns. Tabulated columns will not affect the reading of the tape or performance of the machine during operation.

Another special code is the **end of block code (EOB)** or end of line code (EOL). A **block** of information is one line of information in the source program. A block of information terminates with an EOB code. An EOB code is a solitary hole in track 8. This is the only information contained in track 8 in the RS-244 coding system and tells the machine to execute the previous information. The EOL or EOB character performs two functions: line feed and carriage return. This equates to pushing the RETURN key on the computer.

When you are entering a program into the computer and press the RETURN key, two things happen. The cursor advances one line and returns to the left of the screen (column 1). Pressing the RETURN key while entering a program into the computer outputs an EOB code. When read, this tells the controller to execute the previous block of information.

Another special code is the rewind stop code or end of record (EOR), sometimes placed at the beginning and end of a program. The percent sign (%) is often used as the rewind stop code and end of record code. The end of record code tells the MCU that the reader has read the last piece of information contained on the tape and that it should rewind the tape for the next production run. The rewind stop code is placed at the beginning of the program before any other information. It marks the start of the program and halts the rewind action of the tape reader. These two codes can be found in the first and last blocks of a program, when required.

The DELETE key also produces a special code. The programmer can erase an error with the DEL or DELETE key. This punches a series of holes in the first seven or eight columns of the tape. When the reader or duplicator reads this block, it ignores the block. A tape splice adds new information to the tape, should longer additions be necessary. Usually, the spliced tape is reread through the reader and a new tape made containing the included information. This is because tape readers have a tendency to "eat" spliced tapes, due to the increased thickness at the splice.

When writing the program using the computer, the DELETE key erases the character on the screen. (On a computer, the BACKSPACE key erases typing errors. However, when keypunching a tape, the BACKSPACE key only backs up one space and does not delete the character.) Correcting typing mistakes is much easier on the computer than when keypunching a paper tape. There is a greater chance of recognizing and correcting mistakes before saving a program on the computer.

Now, try to code the statement, N0100 TAB X1 TAB Y5 TAB Z0 EOB using the RS-244 and RS-358 coding systems given in Table 4.3. Include tab codes between words and an EOB code. Did you remember to include parity checking?

4.4 METHODS OF ENTERING NUMBERS

There are four ways of entering information into the MCU: with or without decimal points; no suppression of leading and trailing zeroes; leading zero suppression; and trailing zero suppression. Each line of information in a part program contains one- to eight-bit bytes or **words**. A decimal point is optional in numerical programming, depending on the particular MCU. For some MCUs, zeroes may be omitted, either before a number (leading zeroes) or after a number (trailing zeroes). When it is capable of omitting zeroes, the MCU provides **leading** or **trailing zero suppression.** The method used depends on the make and model of the control unit. The format detail (discussed in Chapter 3) of the MCU is also determined by the make and model of the MCU. The following illustrates the four methods of entering numbers.

1. *Entering a number in every position.* This generally works for all machines. Zeroes are entered before and after the significant digits. Some MCUs require that all numbers be entered without zero suppression.
2. *Leading zero suppression.* The machine reads numbers from right to left and zeroes to the left of the significant digits may be omitted.
3. *Trailing zero suppression.* The machine reads numbers from left to right and zeroes to the right of significant digits may be omitted.
4. *Entering decimal points to define values.*

Decimal Point	No Zero Suppression	Leading Zero Suppression	Trailing Zero Suppression
10.	0100000	100000	010
7.5	0075000	75000	0075
1.	0010000	10000	001
0.5	0005000	5000	0005
0.005	0000050	50	000005

Zero suppression and decimal point programming techniques shorten the coding of position information. If in doubt concerning what kind of suppression a particular MCU uses, enter all of the numbers. All machines will accept this method of entering numbers. Suppression techniques are control-dependent, not machine-dependent. However, you must know the format of values the machine accepts. The format detail and operator's manual of the MCU contain the value formats and suppression techniques available. Format details were discussed in Chapter 3.

4.5 PROGRAMMING FORMATS

The format of the tape specifies the order and arrangement of tape information. The most common tape format is the word address format. The **word address format** uses a letter address to separate each word. A unique letter address identifies each word and minimizes the amount of data coded. The word address format was the first format to use alphanumeric characters to specify data, which provides a more flexible and readable format for coding information. The examples in this text will concentrate on the word address format.

Three other formats that have been used are the fixed sequential, block address, and tab sequential formats. While these formats are obsolete, information on these formats will give you a sense of the developments in information formatting. While studying these formats, keep in mind that each format was developed to provide a simpler and faster method of entering data.

Only numerical data are coded in the fixed sequential format. There are no word address letters to separate words. Coding takes place in a rigid or fixed sequence, with all the necessary codes to control the machine appearing in each block. Each block is the same length and contains the same number of characters. For example, feed and speed data for each movement must appear in each block.

The block address format eliminated the need to include redundant information in subsequent blocks by including a change code. The change code followed the block sequence number (N) and indicated which value(s) changed in relation to the previous block of information.

The **tab sequential format** used the fixed sequential format, while placing tab codes between each word. This made it easier to read the tape and the printout. Tab codes replaced words that were repeated. A tab code preceded each value, and all values were included in each block of information.

FIGURE 4.3
SAMPLE PART PROGRAM SHOWN IN FIGURE 1.2

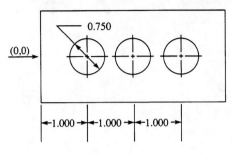

The fixed sequential, block address, and tab sequential formats are **fixed block formats.** A fixed block format does not allow for change in the number or sequence in which words appear within programmed blocks.

One other format available is the interchangeable or compatible format. It is growing in use and acceptance due to its versatility and flexibility. The interchangeable format is basically the same as word address programming, but it permits the use of tab codes to ease readability and has the ability to interchange words within a block of information. This contrasts to other formats in which words are programmed in the order in which they appear in the format detail (fixed block formats). The fixed sequential, tab sequential, word address, and interchangeable tape formats (with and without suppression) are used to code the program given in Fig. 1.2 (shown again in Fig. 4.3).

The format detail used in these examples is

N5 G2 X+34 Y+34 Z+34 I34 J34 K34 F21 S4 T2 M2

where N = sequence number

G = preparatory function (G code)

X,Y,Z,I,J,K = dimension data

F = feed rate

S = speed

T = tool number (and offset)

M = miscellaneous function

(Chapter 3 discusses format details in greater depth.)

Fixed Sequential Format

00001010025000010000000000000000000000000000
00000000001

where the sequence is interpreted as

```
  N    G      X       Y       Z       I       J
00001,01,0025000,0010000,0000000,0000000,0000000,
   K    F    S    T   M
0000000,000,0000,00,01
```

(Commas have been added for readability and would not be coded into the program.)

```
00005,00,0000000,0000000,0000000,0000000,0000000,
0000000,000,0000,00,00
00010,90,-0050000,0000000,0000000,0000000,
0000000,0000000,000,1200,01,06
00020,00,0010000,0000000,0001000,0000000,0000000,
0000000,150,0000,00,00
00030,01,0000000,0000000,-0005000,0000000,
0000000,0000000,000,0000,00,00
00040,01,0000000,0000000,0001000,0000000,0000000,
0000000,000,0000,00,00
00050,00,0020000,0000000,0000000,0000000,0000000,
0000000,000,0000,00,00
00060,01,0000000,0000000,-0005000,0000000,
0000000,0000000,000,0000,00,00
00070,01,0000000,0000000,0001000,0000000,0000000,
0000000,000,0000,00,00
00080,00,0030000,0000000,0000000,0000000,0000000,
0000000,000,0000,00,00
00090,01,0000000,0000000,-0005000,0000000,
0000000,0000000,000,0000,00,00
00100,01,0000000,0000000,0001000,0000000,0000000,
0000000,000,0000,00,00
00110,00,-0050000,0000000,0000000,0000000,
0000000,0000000,000,0000,00,02
```

Note: Words are identified by their location within the block. All words that occur before the last desired word must appear in the block.

Tab Sequential Format

```
     N          G            X              Y        Z    I    J    K    F
 00001 TAB 01 TAB 0007500 TAB 0005000 TAB TAB TAB TAB TAB
     S    T    M
 TAB TAB TAB 01

 00005 TAB 00

 00010 TAB 90 TAB -0050000 TAB 000000 TAB TAB TAB TAB TAB
 TAB 1200 TAB 01 TAB 06

 00020 TAB 00 TAB 0010000 TAB TAB 0001000 TAB TAB TAB TAB
 150

 00030 TAB 01 TAB TAB TAB -0005000

 00040 TAB 01 TAB TAB TAB 0001000

 00050 TAB 00 TAB 0020000

 00060 TAB 01 TAB TAB TAB -0005000

 00070 TAB 01 TAB TAB TAB 0001000

 00080 TAB 00 TAB 0030000

 00090 TAB 01 TAB TAB TAB -0005000

 00100 TAB 01 TAB TAB TAB 0001000

 00110 TAB 00 TAB -0050000 TAB TAB TAB TAB TAB TAB TAB
 TAB TAB 02
```

Note: Words are identified by the number of tab characters preceding the word within the block. If the word has a value of zero or remains unchanged, omit it, but not the tab character.

Word Address Format

```
         N0010G01X0015000Y0010000M06

N0010G00G90X-5Y0S1200T1M6
N0020X1Z.1F15.0
N0030G01Z-.5
N0040Z.1
N0050G00X2
N0060G01Z-.5
N0070Z.1
N0080G00X3
N0090G01Z-.5
N0100Z.1
N0110G00X-5M2
```

Note: Each word has a unique letter address. In this example, decimal points and zero suppression have both been used.

Interchangeable Format

```
N0010 Y0010000 X0015000 M06 G01

N10 G0 G90 X-5. Y0 S1200 T1 M6
N20 Z.1 X1. F15.0
N30 G1 Z-.5
N40 Z.1
N50 G0 X2.
N60 G1 Z-.5
N70 Z.1
N80 G0 X3.
N90 G1 Z-.5
N100 Z.1
N110 G0 X-5. M2
```

Note: In this example, decimal points and zero suppression are both used. In addition, the order of appearance of words may be interchanged.

Advantages and disadvantages of the preceding formats are

1. Fixed sequential requires that all values appear in each block and that each block is the same predetermined length.
2. The block address format eliminates repeating redundant information, but requires the use of a change code to specify which values change between blocks.
3. Tab sequential format uses tab codes to separate data values within the block, but has the same limitations as the fixed sequential format.
4. Word address format is the most popular programming format. Letter addresses identify programmed values, but these addresses must appear in the order specified in the format detail for the MCU.
5. The interchangeable or compatible format is the most versatile format. It also uses letter addresses for programmed values, but does not require any specific order of appearance.

Regardless of the procedure used, each machine uses a particular set of commands and a specific tape format. Machine vocabularies are often very similar, but uniquely different. It is necessary to become familiar with a controller's vocabulary and format detail before beginning any programming.

4.6 Program Transfer Methods

Once the program is written and coded, the coded program must be transferred to the MCU before production begins. One method of program transfer involves reading a previously coded paper tape.

**FIGURE 4.4
ELECTROMECHANICAL READER**

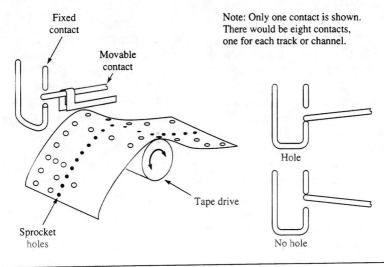

There are two basic types of tape readers: electromechanical and photoelectric. The primary differences between the electromechanical and the photoelectric machines are price and speed. The electromechanical reader is inexpensive and fast enough for cuts that do not require fast access to coded information. These readers run at speeds of 20 to 120 characters per second. They contain eight fixed contacts that ride over the tape. These fixed contacts are part of the control circuitry of the controller and convey programming information to the controller. There are also eight movable contacts, one for each column, which make contact with the fixed contact when sensing coded information. If there is no hole on the tape in the column of a movable contact, this contact touches one side of the fixed contact. If there is a hole, the movable contact drops and contacts the other side of the fixed contact. Depending on which side of the fixed contact the movable contact touches, the MCU receives a 1 or a 0. These contacts can transfer the coded information contained on a tape in a binary form to the MCU. Fig. 4.4 shows an electromechanical tape reader.

When making detailed cuts or several small motions, the controller must read the tape frequently. The electromechanical reader is much too slow for these applications. The solution to this problem is the photoelectric reader. In addition to being faster, the photoelectric reader has a memory. The memory acts as a buffer to hold the previous block, while the reader is reading the next block. As the machine makes a cut, the reader is reading the next block of information into memory. The reader is always one block ahead of the machine. Photoelectric readers are more expensive than electromechanical readers, but they operate at greater speeds, from 100 to 1000 characters per second.

The tape can be friction-fed through the photoelectric reader by a rotating capstan or by a tape-feed sprocket. As the tape feeds through the reader, small exciter lamps shine down onto the tape. Below the tape are nine photodiodes, eight for data and one for timing. A photodiode is a silicon solar cell, similar to half of a transistor. It converts light energy into electric energy like a solar cell, but is much smaller. The small current produced by the photodiode is amplified to increase the signal strength, so that the machine can properly interpret the command.

The eight photodiodes for data lie below the eight columns on the tape. The timing photodiode lies below the sprocket holes and gates the signal from the other circuits. This helps ensure that the reader is interpreting the information correctly, at the proper time. Gating allows the timing of data transfer between the reader and the MCU. For example, the reader will read the tape as a row of information (a byte) passes over the photodiodes. It then stores this information in memory until required by the MCU. As the MCU requires the next block, it signals the reader to send the next block. Picture the gating circuit as the shutter on a camera. When the gating circuit receives a signal from the MCU that new information is required, it opens the shutter and releases the next block of information from memory. Once this information is released, it resets the shutter (gate). Fig. 4.5 shows one type of photoelectric reader.

Tape reader/punch equipment, like the types shown in Figs. 4.4 and 4.5, transfers information entered by the programmer onto tape. This equipment converts the information input by the programmer into rows of holes across the tape. These binary data contain alphanumeric characters, tabs, carriage returns, spaces, and special symbols such as parentheses and asterisks.

Typically, a printed copy of the program verifies the information on the tape. A printed copy is made after punching the tape by reading it through the reader and sending the output to a printer. The tape readback option allows the programmer to check for typing errors and to verify that the tape is an accurate representation of the written program.

After a clean copy of the tape is produced, it may be duplicated any number of times by reading it through the reader fitted with a duplicator or by using a tape duplicator. This creates a backup copy of the tape, just in case the original is damaged.

The following is an introduction to serial communications and is *not* a technical guide to data communications. Detailed information on data communications is contained in a number of computer science textbooks and will not be included here. First, we will define some terms related to data communications used in our discussion of data transfer.

FIGURE 4.5
PHOTOELECTRIC READER

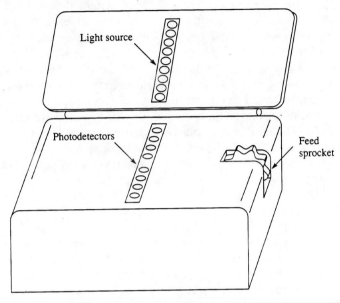

A file is a collection of data stored together. Files may contain text, computer programs, or other information. For example, after a source program is written on the computer, the computer will create a disk file and ask for a file name.

A file name is the title given to a file that uniquely identifies the file. The file name is typically eight characters long and may contain a three-character file extension. For example, "BIRDSEED.DOC" is a valid file name. Notice the period that separates the main file name from the file extension.

A directory is a listing of file names contained on a disk. To list the files contained on a disk, type "dir" and press RETURN. This lists the files in the current logged drive. Floppy disks are typically inserted into internal drives designated "A:" and "B:." To list the files on a floppy disk, type "dir A:" and press RETURN or "dir B:" and press RETURN. The drive light will go on and a listing of the files on that drive will come on the screen.

Computer *hardware* consists of the physical components of the computer system, including the computer, the monitor, and any external physical devices present, such as a mouse, printer, plotter, and other devices. Computer *software* consists of programmed instructions (computer programs) that tell the computer how to do something. Computer games,

CAD/CAM programs, and word processors are all software. Directories generally contain computer software files.

RS–232 is the EIA standard for low-speed serial data communication. Within this standard are a number of parameters concerning voltage levels, timing relationships, and pin diagrams. This technical information goes beyond the scope of this text and is not included here.

What is serial data communication? It is the transferral of data one bit at a time. Communication between the sender and the receiver occurs in a bit stream, with many bits strung together sequentially. In order for the receiver to interpret the information correctly, the receiver must communicate at the same data transfer rate as the sender. The data transfer rate is called the *baud rate*. The baud rate is equal to the number of bits per second (BPS) that is transferred. During serial communication, information is sent through a communications port designated COM on the computer. Most computers have at least one serial port, labeled COM1. Programs are transferred from the computer to the MCU through a COM port in a bit stream at a specified baud rate.

Other parameters associated with serial data communication and program transfer are parity, data bits, stop bits, and duplex. Parity checking for serial data communication is the same as described previously for tape operations. The number of data bits describes the number of bits needed to relay the desired information; there are normally seven or eight data bits. Stop bits are used to define the end of a block of data; there are normally one or two stop bits. Duplex describes the method by which communication occurs. *Full duplex* allows communication to take place in both directions, sender to receiver and receiver to sender. *Half duplex* allows communication in one direction at a time. Consult the manual that accompanies your computer to find detailed information concerning these topics. What is important is that both the computer and the MCU support the same communication parameters. The most common setting is no parity, eight data bits, and one stop bit at full duplex.

The serial port located on the back of the computer allows the computer to interface and talk to other devices. In CNC technology, the computer must communicate with the MCU. Serial data communication using an RS–232 port and an interface cable connection allows the computer to talk to the MCU.

The RS–232 standard covers serial data transfer at rates of 19,200 bits per second and less. Remember that a bit is one piece of information (a 1 or a 0). Program transfer may take place at different rates; however, the sender and receiver of the information must be set for the same transfer, or baud rate. Common baud rates are 300, 1200, 2400, and 9600 BPS. Most controllers fall into this range. Equipment capable of higher speeds is normally capable of operating at the lower speeds also. For example, a machine capable of 2400 BPS would also support 300 and 1200 BPS speeds.

After connecting the hardware and setting the communication parameters on both the computer and the MCU, part program transfer can begin. First, you need to know which communication port is being used. Next, copy the file to this communication port. This transfers the file through the port to the MCU. For example, if you wanted to transfer the file PROGRAM.ONE from a floppy disk in drive A: to the MCU through COM1: (communication port #1), you would follow the following steps.

1. Type "dir A:" to determine if the file is on the disk and to verify the file name.
2. Once you have verified that the file is on drive A:, you would type in "COPY A:PROGRAM.ONE COM1:" followed by a RETURN.
3. The file should then transfer through COM1 to the MCU and the message "[1 file(s) copied]" should appear on the monitor.

This is a simplified explanation of the use of serial data communications to transfer a part program from a floppy disk to the MCU. Now you try it, if you have access to equipment as you read this. Type in the part program listed in Fig. 1.2 using a text editor or word processor. Save the file under the name A:PROGRAM.ONE. Consult the MCU manual to determine the baud rate, parity, number of data bits, number of stop bits, and duplex supported by your MCU. Set the MCU to the desired communication parameters. Then consult your computer's manual and find the section on setting serial communication parameters. Set the computer's parameters the same as the MCU's. Make sure the proper connection exists between the communication ports. Once the hardware and communication parameters have been properly set, follow the previous example and transfer A:PROGRAM.ONE to the MCU.

4.7 SUMMARY

NC/CNC machines use the binary system, which has only two numbers, 1 and 0. Programs fed into NC machines are programmed in binary form, using an appropriate media, code, and format. Programming media include paper tape, magnetic tape, and floppy disks. Coding systems for programming media include the RS-244 (BCD) and RS-358 (ASCII) codes, the most common tape codes used in programming. These codes use seven or eight digits to code information for input into the MCU. In addition, these codes provide for a number of special characters and functions including the parity check, tab codes, EOL or EOB codes, and rewind stop codes or EOR codes. The most common tape format used today is the word address format. Similar to the word address format is the interchangeable, or compatible, format, which is gaining greater acceptance and is even more flexible.

After you enter and code the source program, program transfer can take place. This involves reading the tape through a tape reader or through serial data communications through an interface. There are two basic types of tape punch/reader units: electromechanical and photoelectric. Serial data communications occur through an interface, one bit at a time in a bit stream, which allows direct data transfer between the computer and the MCU. Serial data communications through RS–232 interfaces has become the transfer method of choice.

QUESTIONS AND PROBLEMS

1. List the common tape materials.
2. List the two major types of tape readers and describe the operation of each.
3. What advantages can you see to using floppy disks over other input materials?
4. What disadvantages can you see to using floppy disks?
5. What other materials can you suggest for tape coding?
6. List the two types of tape codes commonly used.
7. What does the term parity mean?
8. How does odd parity checking work?
9. List and describe the major tape formats.
10. Brainstorm five applications for existing or possible technologies that apply to CNC technology as input media.
11. Code your social security number using both the RS–244 and RS–358 coding systems. Remember to use parity checking, include EOB codes, and EOR/Rewind Stop codes.
12. In your own words, define the following terms:
    a. baud rate
    b. data bits
    c. stop bits
    d. duplex
    e. bit stream

Mathematics for Numerical Control Programming

Chapter Objectives

After studying this chapter, the student will be able to
- Describe the application of the law of sines to CNC programming.
- Describe the application of the law of cosines to CNC programming.
- Determine polar and rectangular coordinates.
- Describe the application of trigonometry in solving for cutter offsets.

5.1 Introduction

This chapter will provide an overview of the mathematics used in NC/CNC programming, as well as examples of how to apply these principles to different programming applications. It is often necessary to use mathematics to locate a point or series of points in space. The following information provides a summary of the required mathematics for those unfamiliar with it or as a review for those who may have forgotten some of the material.

5.2 Law of Sines

As an example of the use of mathematics in programming, suppose it was necessary to locate eight equally spaced holes on a 5-in. diameter bolt circle, as shown in Fig. 5.1. Any circle contains 360°. This divided by eight holes is equal to 45° spacing for the holes. This information is used to calculate the angles for each triangle. The length of one of the sides is the radius of the bolt circle, which is one-half the diameter or 2.5 in. Using

FIGURE 5.1
5-IN. BOLT CIRCLE

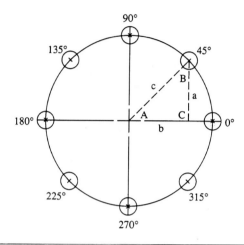

this information and the relationships between the pieces of information, the following calculations may be made:

Angle	X Coordinate		Y Coordinate		X,Y
0°	cos 0° · 2.5 =	2.5 in.	sin 0° · 2.5 =	0 in.	(2.5,0)
45°	cos 45° · 2.5 =	1.77 in.	sin 45° · 2.5 =	1.77 in.	(1.77,1.77)
90°	cos 90° · 2.5 =	0 in.	sin 90° · 2.5 =	2.5 in.	(0,2.5)
135°	cos 135° · 2.5 =	−1.77 in.	sin 135° · 2.5 =	1.77 in.	(−1.77,1.77)
180°	cos 180° · 2.5 =	−2.5 in.	sin 180° · 2.5 =	0 in.	(−2.5,0)
225°	cos 225° · 2.5 =	−1.77 in.	sin 225° · 2.5 =	−1.77 in.	(−1.77,−1.77)
270°	cos 270° · 2.5 =	0 in.	sin 270° · 2.5 =	−2.5 in.	(0,−2.5)
315°	cos 315° · 2.5 =	1.77 in.	sin 315° · 2.5 =	−1.77 in.	(1.77,−1.77)

After you determine the rectangular coordinates (X,Y) of the points required for the bolt circle, the following program may be written:

N100 G0 G90 X-7. Y0 F12. S1800 T1 M6	Proceed to tool/part change location and call tool #1
N110 X-2.5 Z.1	Position over point #1 at clearance plane of .1
N120 G1 Z-.5	Drill to a depth of 0.5
N130 G0 Z.1	Return to clearance plane
N140 X-1.77 Y-1.77	Position over point #2
N150 G1 Z-.5	Drill to depth

N160 G0 Z.1	Return to clearance
N170 X0 Y-2.5	Position over point #3
N180 G1 Z-.5	Drill to depth
N190 G0 Z.1	Return to clearance
N200 X1.77 Y-1.77	Position over point #4
N210 G1 Z-.5	Drill to depth
N220 G0 Z.1	Return to clearance
N230 X2.5 Y0	Position over point #5
N240 G1 Z-.5	Drill to depth
N250 G0 Z.1	Return to clearance
N260 X1.77 Y1.77	Position over point #6
N270 G1 Z-.5	Drill to depth
N280 G0 Z.1	Return to clearance
N290 X0 Y2.5	Position over point #7
N300 G1 Z-.5	Drill to depth
N310 G0 Z.1	Return to clearance
N320 X-1.77 Y1.77	Position over point #8
N330 G1 Z-.5	Drill to depth
N340 G0 Z.1	Return to clearance
N350 X-7. Y0 M2	Return to tool/part change location and end program

Study this example for a moment. Recall from a previous discussion that the code G0 initiates rapid movements and G1 movements at the programmed feed rate. Two miscellaneous functions were used in this example: M6 and M2 (leading zeroes will be suppressed in subsequent examples). These functions perform a tool change and signify the end of the program, respectively. Based on Fig. 5.1 and the previous program written for absolute positioning, rewrite the program using incremental positioning. (The points remain the same, but you must calculate the relative distances between points and use the G91 instead of the G90 code.)

Trigonometry deals with the solution of triangles using various methods. Two pieces of information known about the triangle, such as an angle and the length of one side or the length of two sides, are used to calculate the unknown values.

We will deal primarily with two types of triangles, acute and oblique. The interior angles of acute triangles are all less than or equal to 90°. CNC programming deals primarily with *right triangles,* or those that contain one angle of 90° (a right angle). Right triangles are acute triangles because all of the angles in a right triangle are equal to or less than 90°. Because oblique triangles contain one angle greater than 90°, an oblique triangle cannot be a right triangle. Remember that the sum of the interior angles of any triangle must equal 180°. This chapter deals primarily with right-angle trigonometry.

In addition to two pieces of given information, we must know the relationship between the sides of the triangle and the angles formed by these sides. The terms *opposite, adjacent,* and *hypotenuse* describe the relationship that exists between the angle and the side given. The opposite side of the angle we are dealing with lies directly across from the angle. The adjacent side lies next to the angle. The hypotenuse is always the longest side of the triangle and lies directly across from the largest angle, in a right triangle, the right angle. Referring to Fig. 5.2, with regard to angle A, side a is the opposite side, side b is the adjacent side, and side c is the hypotenuse.

The law of sines describes the relationship that exists between the sine of the angle and length of the side opposite that angle. It states

$$\frac{a}{\sin A} = \frac{b}{\sin B} = \frac{c}{\sin C} \qquad (5.1)$$

From the law of sines we find that (1) the sine of an angle equals the length of the side opposite the angle divided by the length of the hypotenuse; (2) the cosine of an angle equals the length of the side adjacent to the angle divided by the length of the hypotenuse; and (3) the tangent of an angle is equal to the length of the side opposite the angle divided by the length of the side adjacent to the angle. In other words,

$$\sin \Theta = \frac{\text{OPPOSITE}}{\text{HYPOTENUSE}} \qquad (5.2)$$

FIGURE 5.2 □□□□□□□□□□□□□□□□□□□□□□□□□□□
ANGLES AND SIDES OF A RIGHT TRIANGLE

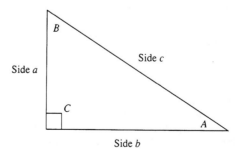

$$\cos \Theta = \frac{\text{ADJACENT}}{\text{HYPOTENUSE}} \qquad (5.3)$$

$$\tan \Theta = \frac{\text{OPPOSITE}}{\text{ADJACENT}} \qquad (5.4)$$

The Pythagorean theorem describes the relationship that exists between the length of the sides of a right triangle: the sum of the squares of the lengths of the two legs of a right triangle equals the square of the length of the hypotenuse, or $c^2 = a^2 + b^2$. (Common triangle notation designates uppercase letters for angles and lowercase letters for sides.) Fig. 5.2 shows the proper designation of the angles and sides of a right triangle.

Using the two known pieces of information (in the left column) and the relationship that exists between the two quantities, the formulas given in Table 5.1 may be used to calculate the unknown quantity. The formulas in Table 5.1 are based on the Pythagorean theorem and the law of sines.

Referring to the bolt circle example, one angle and one side were known. Right-angle trigonometry was used to calculate the length of the remaining sides. The sine function was used to determine the length of the side opposite the angle, or the Y value, and the cosine function was

TABLE 5.1 □□□□□□□□□□□□□□□□□□□□□□□□□□□
RIGHT-ANGLE FORMULAS

Known	Solution		
Side a, Angle A	Side $c = \dfrac{\text{Side } a}{\sin A}$		Side $b = \dfrac{\text{Side } a}{\tan A}$
Side a, Angle B	Side $c = \dfrac{\text{Side } a}{\cos B}$		Side $b = \text{Side } a \cdot \tan B$
Side b, Angle A	Side $c = \dfrac{\text{Side } b}{\cos A}$		Side $a = \text{Side } b \cdot \tan A$
Side b, Angle B	Side $c = \dfrac{\text{Side } b}{\sin B}$		Side $a = \dfrac{\text{Side } b}{\tan B}$
Side c, Angle A	Side $b = \text{Side } c \cdot \cos A$		Side $a = \text{Side } c \cdot \sin A$
Side c, Angle B	Side $b = \text{Side } c \cdot \sin B$		Side $a = \text{Side } c \cdot \cos B$
Sides a and b	Side $c = \sqrt{b^2 + a^2}$		$\tan B = \dfrac{\text{Side } b}{\text{Side } a}$
Sides a and c	Side $b = \sqrt{c^2 - a^2}$		$\sin A = \dfrac{\text{Side } a}{\text{Side } c}$
Sides b and c	Side $a = \sqrt{c^2 - b^2}$		$\sin B = \dfrac{\text{Side } b}{\text{Side } c}$

FIGURE 5.3
STAR EXAMPLE

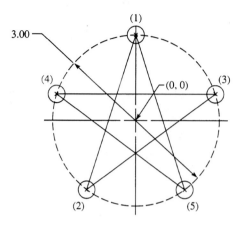

used to calculate the length of the side adjacent to the angle, or the X value. The law of sines may be applied to right triangles where at least an angle and the length of one side or the length of two sides is known.

Another example is given in Fig. 5.3, where an end mill cuts a star-shaped groove into the workpiece. By constructing the triangles shown, the X and Y coordinates are calculated.

Angle	Calculation	Point
18°	cos 18° · 3 = 2.85 in.	(2.85, .93)
	sin 18° · 3 = 0.93 in.	
90°	cos 90° · 3 = 0 in.	(0, 3)
	sin 90° · 3 = 3 in.	
162°	cos 162° · 3 = −2.85 in.	(−2.85, .93)
	sin 162° · 3 = 0.93 in.	
234°	cos 234° · 3 = −1.76 in.	(−1.76, −2.43)
	sin 234° · 3 = −2.43 in.	
306°	cos 306° · 3 = 1.76 in.	(1.76, −2.43)
	sin 306° · 3 = −2.43 in.	

Based on these points, the following incremental program may be written.

N100 G0 G90 X-7. Y0 F12. S1800 T2 M6   Proceed to tool change (TC)/part change (PC) location and call tool #2

N110 X0 Y3. Z.1   Position over point #1 at clearance plane

SECTION 5.3 LAW OF COSINES 109

`N120 G1 G91 Z-.6` Cut into the part surface to an absolute depth of .5 (.6 − .1) in incremental mode

`N130 X-1.76 Y-5.43` Cut to point #2 in the X and Y axes simultaneously

`N140 X4.61 Y3.36` Cut to point #3
`N150 X-5.7 Y0` Cut to point #4
`N160 X4.61 Y-3.36` Cut to point #5
`N170 X-1.76 Y5.43` Cut to point #1
`N170 G0 Z.6` Return to clearance
`N180 G90 X-7. Y0 M2` Return to TC/PC location
End of program

Study this incremental positioning example. Try programming the same star-shaped groove using absolute positioning.

5.3 LAW OF COSINES

The previous discussion concerned right triangles. However, for triangles where none of the angles is a right angle, the calculations are different. Suppose you were given two sides of an oblique triangle and the angle between them or all three sides, what would you do? The law of sines would not apply since there would always be two unknowns. Fig. 5.4 illustrates an example of an oblique triangle with two sides and the angle formed by the two sides known.

The law of cosines states that the square of any side of a triangle equals the sum of the squares of the other two sides minus twice the

FIGURE 5.4 ◻◻◻◻◻◻◻◻◻◻◻◻◻◻◻◻◻◻◻◻◻◻◻◻◻◻
OBLIQUE TRIANGLE WITH TWO SIDES AND THE INCLUDED ANGLE KNOWN

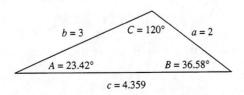

product of the two known sides and the cosine of the angle formed by the two known sides.

$$a^2 = b^2 + c^2 - 2bc(\cos A), \tag{5.5}$$

$$b^2 = a^2 + c^2 - 2ac(\cos B), \text{ and} \tag{5.6}$$

$$c^2 = a^2 + b^2 - 2ab(\cos C). \tag{5.7}$$

Suppose that the two sides were 2 and 3 (a and b) and the included angle (C) was 120°. The third side would equal

$$c^2 = a^2 + b^2 - 2ab(\cos C)$$

$$c^2 = 2^2 + 3^2 - 2(2)(3)(\cos 120)$$

$$c^2 = 4 + 9 - (-6) = 19$$

Therefore, $c = 4.359$.

Recall the law of sines,

$$\frac{a}{\sin A} = \frac{b}{\sin B} = \frac{c}{\sin C}$$

From the law of sines,

$$\frac{a}{\sin A} = \frac{c}{\sin C}$$

so

$$\frac{a \sin C}{c} = \sin A$$

Filling in the numbers we get

$$\sin A = \frac{2 \cdot \sin 120}{4.359} = 0.3974$$

and

$$\sin^{-1}(0.3974) = 23.42°$$

Since the sum of the interior angles of a triangle must equal 180°, $B = 180° - (120° + 23.42°) = 36.58°$. In summary, side $a = 2$, $b = 3$, $c = 4.359$, angle $A = 23.42°$, $B = 36.58°$, and $C = 120°$. The following table summarizes the calculations for oblique triangles (see Fig. 5.5).

FIGURE 5.5 □□□□□□□□□□□□□□□□□□□□□□□□□□
OBLIQUE TRIANGLE

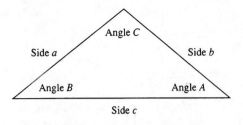

Known	Solutions			
One side and two angles (For example, a, A, and B)	Side $b = \dfrac{\text{Side } a \cdot \sin B}{\sin A}$		Side $c = \dfrac{\text{Side } a \cdot \sin C}{\sin A}$	
Two sides and the included angle (For example, a, b, and C)	$\tan A = \dfrac{\text{Side } a \cdot \sin C}{\text{Side } b - (\text{Side } a \cdot \cos C)}$		Side $c = \dfrac{\text{Side } a \cdot \sin C}{\sin A}$	
Two sides and angle opposite one side (For example, a, b, and A)	$\sin B = \dfrac{\text{Side } b \cdot \sin A}{\text{Side } a}$		Side $c = \dfrac{\text{Side } a \cdot \sin C}{\sin A}$	
All three sides: a, b, and c	$\cos A = \dfrac{b^2 + c^2 - a^2}{2bc}$		$\sin B = \dfrac{\text{Side } b \cdot \sin A}{\text{Side } a}$	

Let's take another example where we know all three sides and none of the angles. We want to find all three angles in an oblique triangle where side $a = 10$, $b = 18$, and $c = 15$. From the law of cosines we can derive the formula

$$\cos A = \frac{b^2 + c^2 - a^2}{2bc}$$

Substituting the numbers in,

$$\cos A = \frac{18^2 + 15^2 - 10^2}{2(18)(15)} = \frac{449}{540} = 0.8315 \quad \cos^{-1}(0.8315) = 33.75°$$

$$\cos B = \frac{a^2 + c^2 - b^2}{2ac} = \frac{10^2 + 15^2 - 18^2}{2(10)(15)} = \frac{1}{300} = 0.0033$$

$$\cos^{-1}(0.0033) = 89.81°$$

Therefore, $C = 180° - (A + B) = 180° - (33.75° + 89.81°) = 56.44°$.

These examples give you some concept of the importance of mathematics (and of owning a calculator with trigonometric functions!) in CNC programming.

5.4 POLAR NOTATION

In addition to the rectangular coordinates (discussed in Chapter 3), points may be designated as polar coordinates. The **polar coordinate system** requires a magnitude and a direction, e.g., 5 cm at 60°. Consider the bolt circle example given in Fig. 5.1. Each of the eight holes could be designated in polar form: 2.5 in. at 0°, 2.5 in. at 45°, 2.5 in. at 90°, 2.5 in. at 135°, 2.5 in. at 180°, 2.5 in. at 225°, 2.5 in. at 270°, and 2.5 in. at 315°. The radius of the bolt circle is 2.5 in., which specifies the magnitude of each of the vectors. A vector is a straight line that has a magnitude or length and a direction. Vectors point in only one direction. The vectors 2.5 in. at 0° and 2.5 in. at 180° have the same magnitude, but point in opposite directions.

Looking at Fig. 5.1 and using the polar coordinates of the points, we can calculate the rectangular coordinates (X and Y values) for these points. The following table illustrates the calculations of rectangular coordinates in Fig. 5.6 based on the polar form given using the sine and cosine functions.

Point	Polar Form	X (in.)	Y (in.)
1	2.5∠0°	2.5(cos 0°) = 2.5	2.5(sin 0°) = 0
2	2.5∠45°	2.5(cos 45°) = 1.77	2.5(sin 45°) = 1.77
3	2.5∠90°	2.5(cos 90°) = 0	2.5(sin 90°) = 2.5
4	2.5∠135°	2.5(cos 135°) = −1.77	2.5(sin 135°) = 1.77
5	2.5∠180°	2.5(cos 180°) = −2.5	2.5(sin 180°) = 0
6	2.5∠225°	2.5(cos 225°) = −1.77	2.5(sin 225°) = −1.77
7	2.5∠270°	2.5(cos 270°) = 0	2.5(sin 270°) = −2.5
8	2.5∠315°	2.5(cos 315°) = 1.77	2.5(sin 315°) = −1.77

Try this procedure using a bolt circle with a 3-in. radius. Create a table as in the example and find the rectangular coordinates of the eight hole locations.

We have discussed both rectangular and polar notations for point locations. Now we will translate rectangular coordinates into polar coordinates. The previous examples and exercise worked through going from polar coordinates to rectangular notation. Look at the following example of converting from rectangular coordinates to polar notation.

FIGURE 5.6
POLAR COORDINATES

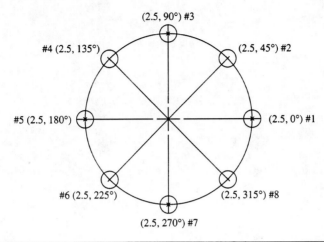

The following table of rectangular coordinates is to be converted to polar form. The conversion hinges on the concepts that the X value is the side adjacent to the angle and the Y value is the side opposite the angle we are looking for to convert to polar form. Therefore, we can use the tangent function, where $\tan A = Y/X$. This gives us the angle formed by the two sides. The length of the hypotenuse is found using the Pythagorean theorem, $c^2 = a^2 + b^2$, where the lengths of sides a and b are determined from the rectangular coordinates.

Point	X (in.)	Y (in.)	c (in.)	A
1	1.5	0	1.5	0°
2	0.75	1.3	1.5	60°
3	−0.75	1.3	1.5	120°
4	−1.5	0	1.5	180°
5	−0.75	−1.3	1.5	240°
6	0.75	−1.3	1.5	300°

These six holes lay on a 3 in.-diameter bolt circle. In NC/CNC programming, rectangular coordinates are generally programmed. However, part features may be specified using polar notation. Try converting the following table of rectangular coordinates to polar form.

Point	X (in.)	Y (in.)
1	2.60	1.50
2	2.12	2.12
3	0.78	2.90
4	−1.27	2.72
5	−2.12	2.12
6	−2.95	−0.52
7	−2.12	−2.12
8	−1.27	−2.72
9	1.50	−2.60
10	2.90	−1.50

Each of these examples referenced a central origin, or pole, on which the magnitudes and angular values were based. This pole or origin is typically at (0,0). Study these examples and try to plot their solutions. Write a program to drill the set of holes given in the previous exercise. Use a Z depth of −0.5 and both absolute and incremental positioning.

5.5 Calculating Cutter Offsets

The following section deals with the calculation of **cutter offsets** for linear and circular interpolation. Modern controllers generally offer tool radius/diameter compensation, which calculates the cutter offset and properly positions the cutting tool automatically. Tool radius/diameter compensation was discussed in Chapter 2. Manual calculation of cutter offsets is unnecessary if the MCU has the ability to compensate automatically.

Trigonometry is commonly used to calculate cutter offsets in CNC programming. Offsets are used with linear or circular interpolation to position the cutter prior to machining. **Linear interpolation** is the simultaneous control of the various axes to produce a straight-line cut at any angle. The MCU calculates the angle necessary to connect the programmed points. Based on the slope of the line between the current tool position and the next programmed point, the MCU proportionately controls the axis drive motors in relation to the slope ratio. **Circular interpolation** is used to cut arcs and arc segments. Linear and circular interpolations are programmed on both turning and milling machines and may be performed in three or more axes, depending on the controller.

Cutter offsets are added or subtracted from the programmed part coordinates, depending on the direction of the cut. Without cutter offsets, part features will be greater or less than desired. Cutter offsets are used to position the cutter properly to prevent undercuts or leaving excess material due to the angles or arcs involved. Fig. 5.7 provides a representation of the need for cutter offsets in linear and circular interpolation.

As an example of the calculations involved in determining the cutter offsets for linear interpolation, look at Fig. 5.8. The chamfer shown is for

FIGURE 5.7
REASONS FOR PROGRAMMING CUTTER OFFSETS

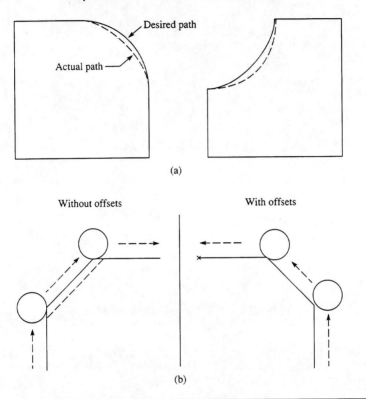

the lathe. However, the calculations are the same for the mill; only the axes will change. Notice the tool must be positioned away from the part in both the X and Z axes. These are the cutter offsets. Calculated and programmed cutter locations include cutter offsets a known distance from the part feature.

Fig. 5.9 is an example of calculating cutter offsets for chamfer programming on the lathe.

If the part required a 0.050 × 30° chamfer, using a 0.040 tool nose radius (TNR) tool, the X and Z offsets are

$$\tan 0° = \text{Opposite/Adjacent}$$

$$\tan 30° = \frac{\text{Offset}_z}{0.040}$$

FIGURE 5.8 □□□□□□□□□□□□□□□□□□□□□□□□□
CUTTER OFFSETS FOR CHAMFER PROGRAMMING

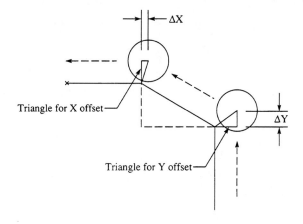

$$\text{Offset}_z = \tan 30° \cdot 0.040 = 0.577 \cdot 0.040 = 0.023 \text{ in.}$$

$$\tan 15° = \frac{\text{Offset}_x}{0.040}$$

$$\text{Offset}_x = \tan 15° \cdot 0.040 = 0.268 \cdot 0.040 = 0.011 \text{ in.}$$

FIGURE 5.9 □□□□□□□□□□□□□□□□□□□□□□□□□
CUTTER OFFSETS FOR CHAMFER PROGRAMMING ON A LATHE

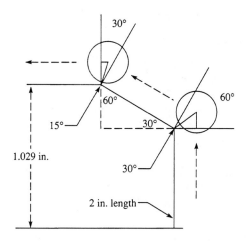

When programming the chamfer, the X-offset value (0.011 in.) is subtracted from the start point radius value (X), and the length value (Z) remains the same as on the part drawing. The Z-offset value (0.023 in.) is added to the end point Z value, and the X value (finished radius) remains the same.

This may seem confusing, but take several moments to look at the following programming statements. Examine the program steps and determine the effect of the offsets. Refer to Fig. 5.9 for the specified lengths and diameters.

G0 G90 X.989 Z2. F120 S2500 T0101   Places the cutter at the start point with the X-offset value entered

G1 X1.029 Z1.973   Cuts the chamfer ending at the final diameter with the required Z offset entered

The previous example showed the offset calculations used with linear interpolation on a lathe or turning center. The next example is for a milling machine or machining center. Fig. 5.10 shows the part feature for the following example.

Using a 0.5-in. diameter cutter, the following calculations are made to find the X and Y offsets to cut a 0.050 in. deep × 30° angular cut.

$$\tan 15° = \frac{\text{Offset}_X}{0.250}$$

$$\text{Offset}_X = \tan 15° \cdot 0.250 = 0.268 \cdot 0.250 = 0.067 \text{ in.}$$

$$\tan 30° = \frac{\text{Offset}_Y}{0.250}$$

$$\text{Offset}_Y = \tan 30° \cdot 0.250 = 0.577 \cdot 0.250 = 0.144 \text{ in.}$$

Take several moments to look at the following programming statements. Examine the program blocks to determine the effect of the cutter offsets. Refer to Fig. 5.10 for the specified part feature.

G0 G90 X2. Y1.856 F12 S2500 T1   Places the cutter at the start point with the Y-offset value entered

G1 X1.883 Y2.029   Cuts the chamfer ending at the end point with the required X offset entered

FIGURE 5.10
MILLING EXAMPLE OF CUTTER OFFSETS

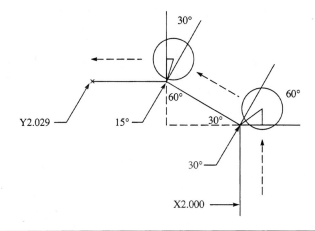

Circular interpolation is the ability to cut **arcs** or arc segments. CNC machines without circular interpolation ability are severely limited in use. Most modern CNC machines offer circular interpolation capability.

The MCU generates angular chord segments to approximate arcs and arc segments. These segments are necessary since the cutter does not revolve around a fixed centerpoint. An MCU equipped with circular interpolation ability calculates and cuts a series of chord segments to approximate the programmed arc. These chord segments are practically indistinguishable from the true arc. Fig. 5.11 compares the actual cutter path (chord segments) with the true arc. As you can see, the more segments cut, the closer the arc segments approximate the true arc.

Circular interpolation is programmed using one of two methods in word address format. Regardless of the method used, the G2 and G3 codes cause the MCU to initiate cutter motion in the clockwise and counterclockwise directions, respectively. This section will briefly introduce circular interpolation; greater detail is in Chapter 6.

In addition to the codes G2 and G3, circular interpolation uses the I, J, K, and R words. I, J, and K define the X, Y, and Z coordinates of the arc centerpoint. The **R address** defines the arc radius. The first method of circular interpolation (the centerpoint method) involves (1) the appropriate G code; (2) the coordinates of the arc endpoint; and (3) the distance from the current tool position to the arc centerpoint. For example, G2 X0 Y−1.5 I1.5 J0 would cut a clockwise arc beginning at (X1.5, Y0) defined by the I and J coordinates. The arc would end at (X0, Y−1.5). This statement cuts a full quadrant.

We can program arc segments in the same way. Suppose we wanted to cut half a quadrant using the same radius. First, we need to calculate

FIGURE 5.11
ACTUAL VERSUS DESIRED CUTTER PATH

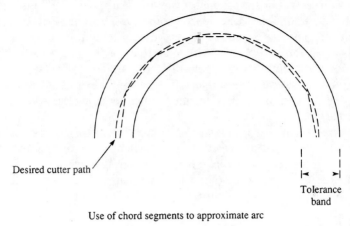

Use of chord segments to approximate arc

---

the endpoint coordinates of the segment. We know the radius (1.5) and the angle (45°), so we can use trigonometry to calculate the coordinates. The following calculations illustrate the procedure used.

$$X = 1.5 \cdot \cos 45° = 1.061 \qquad Y = 1.5 \cdot \sin 45° = 1.061$$

Using the calculated endpoint coordinates and the centerpoint coordinates already given, the following block of information is programmed: G2 X1.061 Y1.061 I1.5 J0. Some MCUs allow single-statement programming of arcs up to 90°, where the arc must be contained in one quadrant. Circular interpolation in these circumstances requires four statements to cut a full 360° circle. Other MCUs allow 360° arcs in a single statement. Refer to the machine's operations manual to determine the capabilities of the MCU you will be using.

Lathe programming uses the same procedure. To cut a similar arc on the lathe using the same information given in the previous example, use G2 X1.061 Z1.061 I1.5 K0. The difference is the programmed dimension words.

The second method used for circular interpolation (the radius method) uses the radius of the arc and the R address rather than programmed centerpoint coordinates. Compare the two methods:

Centerpoint

G2 X0 Y-1.5 I1.5 J0
G2 X1.061 Y1.061 I1.5 J0
G2 X1.061 Z1.061 I1.5 K0

Radius

G2 X0 Y-1.5 R1.5
G2 X1.061 Y1.061 R1.5
G2 X1.061 Z1.061 R1.5

The two methods, centerpoint and radius, are used to program circular interpolated arcs and arc segments in word address format. When the arc intersects a linear segment or another arc, cutter offsets must be calculated to maintain proper part geometry. The following discussion will center on calculating cutter offsets in these situations.

Figs. 5.12 through 5.15 provide examples of the use of cutter offset calculations in CNC programming. These standard formulas provide a shortcut to calculating cutter offsets. They allow substituting specific program information into standard formulas, thus reducing the necessary offset calculation.

We will use these formulas to calculate the cutter offsets necessary to program an arc that intersects a line parallel to an axis and an arc that intersects another arc. We will be using a 0.5-in. diameter end mill cutter.

Consider the example in Fig. 5.16, where an arc intersects a line parallel to the X axis. First, we must calculate the differences in I and J.

$$\Delta J = 1 - .75 = .25$$

Now that we know two sides of the triangle shown in Fig. 5.16, we can use the Pythagorean theorem to calculate the missing side.

$$\Delta I = \sqrt{(1^2 - .25^2)} = 0.9682$$

Using the formulas in Fig. 5.14, the following calculations for the X and Y arc endpoint coordinates are performed:

$$\Delta X = \Delta I - \sqrt{(R - CR)^2 - (\Delta J - CR)^2}$$

$$\Delta X = 0.9682 - \sqrt{0.5625}$$

$$\Delta X = 0.2182$$

$$\Delta Y = CR \text{ (Cutter Radius)} = 0.25$$

Using the same formulas in Fig. 5.14, the following calculations for the X and Y arc starting points are performed.

$$\Delta X = CR = 0.25$$

$$\Delta Y = \Delta J - \sqrt{(R - CR)^2 - (\Delta I - CR)^2}$$

$$\Delta Y = 0.25 - \sqrt{0.0466}$$

$$\Delta Y = 0.0339$$

Now that we know the X- and Y-offset values, we can add them to our program.

```
G0 G90 X3.25 Y2.0339    X = 3 + 0.25        Y = 2 + 0.0339
G2 X2.4682 Y3. R1.      X = 2.25 + 0.2182   Y = 2.75 + 0.25
```

FIGURE 5.12 OFFSET CALCULATIONS: INTERSECTING LINES NOT PARALLEL TO AN AXIS

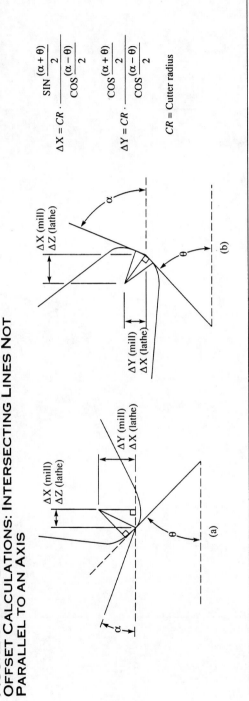

$$\Delta X = CR \cdot \frac{\text{SIN}\frac{(\alpha+\theta)}{2}}{\text{COS}\frac{(\alpha-\theta)}{2}}$$

$$\Delta Y = CR \cdot \frac{\text{COS}\frac{(\alpha+\theta)}{2}}{\text{COS}\frac{(\alpha-\theta)}{2}}$$

CR = Cutter radius

FIGURE 5.13
OFFSET CALCULATIONS: LINES TANGENT TO CURVES NOT PARALLEL TO AXIS

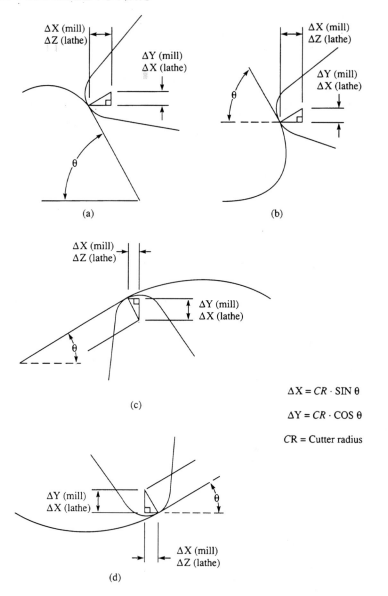

$\Delta X = CR \cdot \text{SIN } \theta$

$\Delta Y = CR \cdot \text{COS } \theta$

CR = Cutter radius

FIGURE 5.14 ◻◻◻◻◻◻◻◻◻◻◻◻◻◻◻◻◻◻◻◻◻◻◻◻◻◻
OFFSET CALCULATIONS: INTERSECTION OF LINE AND CIRCLE PARALLEL TO AXIS

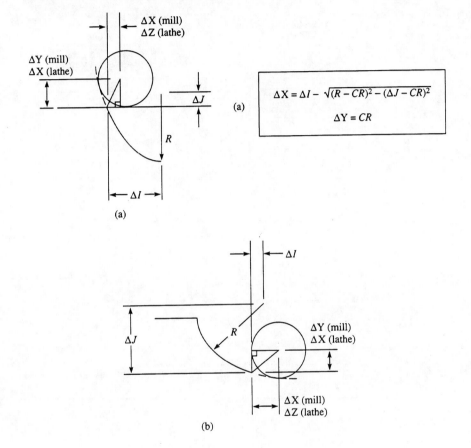

$$\Delta X = \Delta I - \sqrt{(R-CR)^2 - (\Delta J - CR)^2}$$
$$\Delta Y = CR$$

(a)

$$\Delta X = CR$$
$$\Delta Y = \Delta J - \sqrt{(R-CR)^2 - (\Delta I - CR)^2}$$

(b)

CR = Cutter radius

FIGURE 5.15
OFFSET CALCULATIONS: INTERSECTION OF CIRCLE AND LINE PARALLEL TO AXIS

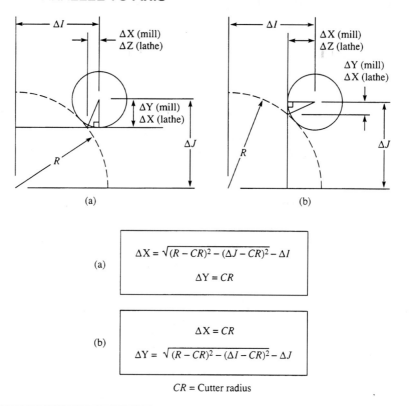

(a)
$$\Delta X = \sqrt{(R - CR)^2 - (\Delta J - CR)^2} - \Delta I$$
$$\Delta Y = CR$$

(b)
$$\Delta X = CR$$
$$\Delta Y = \sqrt{(R - CR)^2 - (\Delta I - CR)^2} - \Delta J$$

CR = Cutter radius

Now consider an arc that intersects a linear segment. Referring to Fig. 5.17 and Fig. 5.13, the following calculations are made.

$$\Delta X = CR \cdot \sin 30° = 0.5 \cdot 0.5 = 0.25$$
$$\Delta Y = CR \cdot \cos 30° = 0.5 \cdot 0.8660 = 0.2165$$

As in the previous example, these offset values are added to the programmed values to maintain the correct cutter path. For example,

```
G1 G90 (X + 0.25) (Y + 0.2165)
G3 X. . . . Y. . . . R. . . .
```

FIGURE 5.16
ARC INTERSECTION EXAMPLE

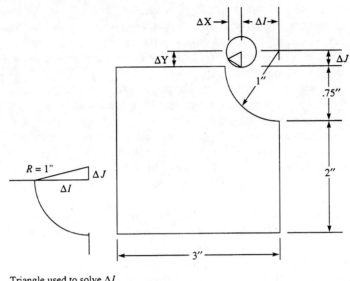

Triangle used to solve ΔI

FIGURE 5.17
ARC INTERSECTION EXAMPLE

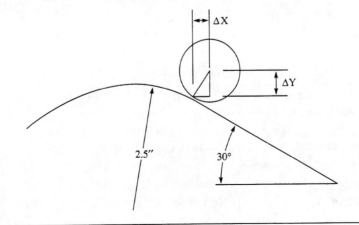

5.6 Summary

Trigonometry is used to solve for the unknowns in triangles when appropriate information about the triangle is known. The law of sines and the law of cosines are used to calculate unknowns from the information given based on the relationship that exists between the information supplied. The Pythagorean theorem relates the relationship that exists between the sides of a right triangle.

Mathematics is important in calculating programmed points and cutter offsets. Cutter offsets are programmed to maintain part geometry in linear and circular interpolation. Linear interpolation is the MCU's ability to cut straight-line segments by controlling one or more axis drive motors proportionately. Control is based on the ratio formed by the slope of the line between the current cutter position and the next programmed point. Circular interpolation is the ability of the MCU to cut arcs and arc segments by cutting small chord segments.

There are two primary methods of circular interpolation in the word address format, centerpoint and radius. The centerpoint method uses (1) the appropriate G code; (2) the arc endpoint coordinates; and (3) the distance from the current tool position to the arc centerpoint. The format for the centerpoint method is

G2/G3 X. . . . Y. . . . (Z. . . .) I. . . . J. . . . (K. . . .).

The second method of circular interpolation is the radius method. The radius replaces the arc centerpoint definition in the radius method. It uses (1) the appropriate G code; (2) the arc endpoint coordinates; and (3) the radius of the programmed arc or arc segment. The format for the radius method is

G2/G3 X. . . . Y. . . . (Z. . . .) R. . . .

in word address format.

Questions and Problems

1. How are the sine, cosine, and tangent defined for angles in a right triangle using trigonometry?
2. Construct a right triangle having sides of 3 in., 4 in., and 5 in. and calculate the three included angles using trigonometry.
3. What would the angles for the triangle in Problem #2 be if the length of the sides were doubled?
4. Calculate the X and Y coordinates of the five holes shown in Fig. 5.18.

**FIGURE 5.18
PROBLEM #4**

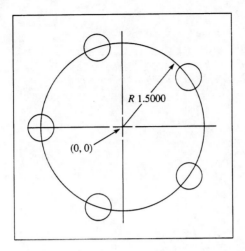

5. Calculate the X- and Z-offset values for a 0.100 × 45° chamfer, using a 0.042 TNR on the lathe.
6. Write the program statements necessary to cut the chamfer in Problem #5 if the finish diameter is 2 in. and starting Z coordinate is zero. Program the cutter centerline.

**FIGURE 5.19
PROBLEM #11**

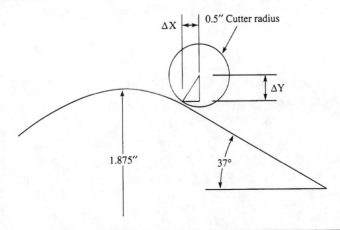

**FIGURE 5.20
PROBLEM #12**

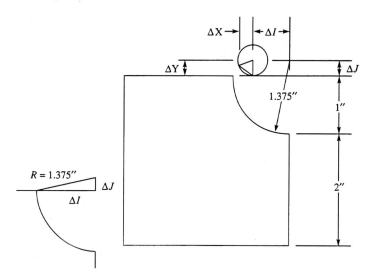

Triangle used to solve Δ*I*

7. Calculate the X and Y offset values for a 0.25 in. × 30° chamfer using a 0.375 in.-diameter end mill cutter.
8. Write the program statements necessary to cut the chamfer in Problem #7 if the finish Y coordinate is zero and the starting X coordinate is 1.5. Program the cutter centerline.
9. Define the centerpoint method of circular interpolation. Give the format for the centerpoint method.
10. Define the radius method of circular interpolation. Give the format for the radius method.
11. Calculate the X- and Y-offset values for the arc given in Fig. 5.19.
12. Calculate the offset values for the segment in Fig. 5.20.

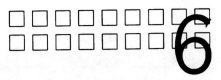

Linear, Circular, and Helical Interpolation

Chapter Objectives

After studying this chapter, the student will be able to
- Recognize and define linear interpolation.
- Describe the procedures for programming part features using linear interpolation.
- Calculate linear approximations of arcs.
- Recognize and define circular interpolation.
- Program part features using radius and centerpoint circular interpolation methods.
- Describe the procedures and program part features in the alternate plane pairs.
- Recognize and define helical interpolation.
- Describe the procedures and program part features using helical interpolation.

6.1 Introduction

Linear interpolation is the ability to cut straight, angular segments beginning at one point and ending at another, using a NC/CNC machine. This takes place in any of the quadrants, at any angle. Linear interpolation may be used to approximate a circular entity by making a series of smaller, straight-line cuts within the boundary tolerances of the circle. Circular interpolation is the ability to cut arcs, partial arcs, or full circles using numerical control programming. Circular interpolation uses simultaneous control of two or more axes.

Helical interpolation allows circular interpolation in two axes and linear control in a third axis. Helical interpolation is used to program helixes. These features are among the more important capabilities of a CNC machine. This chapter will describe linear, circular, and helical interpolation procedures, formats, and capabilities.

6.2 LINEAR INTERPOLATION

When performing straight-cut linear interpolation, the MCU controls two or more machine axes simultaneously. In cutting angular segments, the controller uses the programmed information to calculate the angle of cut, or the slope, of the linear segment. The change in length over the distance from starting point to ending point in the axes can be used to determine the slope of the linear segment. The ratio determined by the slope determines the drive motor speeds for the axes. For example, if the slope of the segment were one-half, or, in other words, if the ratio is 1:2, the drive motor would move the tool twice as fast in one axis as in the other axis. Fig. 6.1 provides an example of linear interpolation for a rectangular frame.

The program required to cut the groove shown in Fig. 6.1 using a 0.25-in. cutter is

```
N10 G0 G90 X-4. Y0 T1 M6         Tool change (position #1)
N20 X.625 Y.625 F12. S2000       Position #2
N30 G1 Z-.125                    Feed to depth
N40 X4.375                       Feed to position #3
N50 Y2.375                       Feed to position #4
N60 X.625                        Feed to position #5
N70 Y.625                        Feed to position #6
N80 G0 Z.1                       Rapid to clearance plane
N90 X-4. Y0 M2                   Rapid return to TC/PC location
                                 End of program
```

Linear interpolation can be performed in any one of the plane pairs, X-Y, Z-X, Y-Z, or in all three axes simultaneously.

Another use for linear interpolation is in the approximation of curves, arcs, or circles. This use is generally limited to machines that are not capable of circular interpolation. Fig. 6.2 is an example of the linear approximation of a circle using straight-line segments; first a square is used, second, a hexagon, and finally an octagon is enclosed by a circle. Each of these segments forms a chord, secant, or tangent of the circle. The more segments cut, the closer the approximation comes to the circle. The number and length of the programmed segments required depend on the maximum allowable tolerance for the deviation from the circle's true shape.

FIGURE 6.1
LINEAR INTERPOLATION

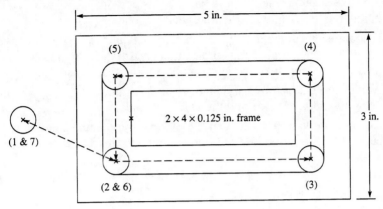

As the maximum allowable tolerance for the approximation becomes smaller, more programmed segments are needed for the approximation to remain within tolerance. The more segments programmed, the longer the program's length and the longer the execution time for the program. This is due to the machine having to read, interpret, and cut a short distance, then read, interpret, and cut a short distance, and so on until the arc is complete.

As an introduction to circular interpolation, consider the programming steps necessary on machines that do not have circular interpolation capabilities. In this situation, arcs and arc segments must be programmed using linear segments. Each of these linear cuts must stay within the upper and lower tolerance bands of the arc or segment. The following procedure is used to program a linear approximation of an arc.

STEPS USED IN CALCULATING LINEAR INTERPOLATED ARCS

1. Adjust the tolerance of the arc to correspond with the machine's accuracy.

FIGURE 6.2
LINEAR APPROXIMATIONS OF A CIRCLE

2. Calculate the angle produced by the hypotenuse and adjacent side of the triangle. The hypotenuse is equal to the arc radius plus the upper limit of the tolerance. The adjacent side is equal to the arc radius minus the lower limit of the tolerance.
3. Calculate the number of times that the calculated angle will divide into 90°. If the number is a whole number, use it. If the number is not a whole number, take the next smallest number and use it.
4. Adjust the angle to satisfy Step #3 and recalculate the hypotenuse using the adjacent side and the adjusted angle.
5. Average the variance between the original hypotenuse and the recalculated hypotenuse. Add half the variance to the recalculated hypotenuse.
6. Using the hypotenuse found in Step #5 and the adjusted angle, recalculate the adjacent side.
7. The first point on the arc will be the length of the adjacent side on the X axis and zero, if the centerpoint of the arc is the origin (0,0).
8. Successive points are found by adding a full or half angle to the angle used in Step #6 and recalculating the lengths of the adjacent sides (X values) and opposite sides (Y values). The hypotenuse for these calculations is the one found in Step #4.

EXAMPLE

Given that X and Y zero is located at the center of the arc. The arc radius is 0.5 in. The desired tolerance is +/−0.002 in.

1. Machine tolerance can be ignored for the purposes of this example. Tolerance is +/−0.002 in.
2. 0.5 in. +/−0.002 in. = 0.502/0.498. The hypotenuse is equal to 0.502 in. and the adjacent side is equal to 0.498 in. The calculated angle is $\cos^{-1}\dfrac{0.498}{0.502} = 7.237°$.
3. $\dfrac{90°}{7.237°} = 12.43$. Next smallest angle is 6°.
4. Using 6° as the angle, $\dfrac{0.498}{\cos 6°} = 0.5007$ (new hypotenuse).
5. $0.502 - 0.5007 = 0.0013$. $\dfrac{0.0013}{2} = 0.00065$.
   $0.5007 - 0.00065 = 0.50005$.
6. Using the hypotenuse of 0.50005, the adjacent side is equal to $\cos 6° \cdot 0.50005 = 0.4973$.
7. Point 1 is equal to (0.4973,0).

8. The following table gives the programmed points for full-angle interpolations of the arc.

Point	Angle (°)	X (in.)	Y (in.)
1	0	0.4973	0
2	6	0.4980	0.0523
3	12	0.4898	0.1041
4	18	0.4762	0.1547
5	24	0.4574	0.2037
6	30	0.4336	0.2504
7	36	0.4051	0.2943
8	42	0.3721	0.3350
9	48	0.3350	0.3721
10	54	0.2943	0.4051
11	60	0.2504	0.4336
12	66	0.2037	0.4574
13	72	0.1547	0.4762
14	78	0.1041	0.4890
15	84	0.0523	0.4980
16	90	0	0.4973

As you can see, this method works but is slow and tedious. There is a better way to program arcs and arc segments—circular interpolation. Machines equipped with circular interpolation capabilities speed up and simplify the programming of arcs and arc segments by allowing the machine to calculate the cutter positions for you.

6.3 CIRCULAR INTERPOLATION

Circular interpolation is the ability to cut a circular path, varying from a small arc segment to a full circle. The controller generates the cutter path based on the programmed information. Circular interpolation uses significantly fewer programming steps and less execution time than the same linear approximation.

Most modern controllers allow the programming of arcs from 1 to 360° in whole-angle increments. Some controllers may be limited to the programming of arcs 90° or less that are contained in the same quadrant. Arcs of greater than 90° or which cross quadrant boundaries would require more than one block of information. For controllers with limited circular capabilities, arcs of 90° or less can be programmed in one block, while four blocks of information (one quadrant at a time) are required in programming a full circle. The number of blocks required depends on the type of circular interpolation used and the capabilities of the MCU.

Four pieces of information are necessary to program an arc in circular interpolation:

1. The direction of travel (CW or CCW)

2. The starting point of the arc
3. The ending point of the arc
4. The center of the arc

The EIA standard (RS–274D) G codes for circular interpolation are G2 and G3, for clockwise (CW) and counterclockwise (CCW) travel, respectively. These codes initiate cutter motion in the direction indicated from the positive end of the axis perpendicular to the plane of interpolation. For example, the CW and CCW directions are specified from the positive Z axis for the X-Y plane pair.

Circular interpolation is initiated by positioning the cutter at the starting point of the arc. The center point of the arc is known and the distance from the starting point of the arc to the center of the arc is programmed using the secondary words I, J, and K for the X, Y, and Z coordinates on milling machines, respectively. I and K are used to designate the center for the X and Z axes on turning machines.

Circular interpolated arcs may be cut in any of the plane pairs. The codes G17, G18, and G19 are used to select the plane in which functions such as circular interpolation and cutter compensation are performed. G17 selects the X-Y plane, G18 selects the Z-X plane, and G19 selects the Y-Z plane. Usually, I, J, and K are absolute values regardless of whether programming is done in the absolute or incremental mode. The programmed center point is modal in the absolute mode, but must be programmed in each block for incremental programming.

The ending point of the arc is the final resting point of the cutter when the arc is finished and is programmed using the X, Y, and Z words. The ending point is nonmodal and must be programmed in each block. When programming arcs of greater than 90°, the point at which the arc crosses from one quadrant to another must be programmed as the ending point. The controller assumes that this programmed end point is the starting point for the next arc. The next programmed block only requires a new ending point, unless the center point has changed. If the programmed end point does not fall on the programmed arc, cutter motion is unpredictable. Typically, no program error is given and the error is not recognized until the cutter deviates from the intended path. This is one reason why all programs should be dry-run first.

The following program illustrates the steps in circular interpolation using the X, Y, I, and J words in Fig. 6.3.

The programming steps required to cut the arc shown in Fig. 6.3 using X, Y, I, and J at a depth of 0.25 in. are

```
N10 G0 G90 X-4. Y0 S2000 T1 M6     Rapid to TC/PC
N20 X0 Y0 Z.1 F12.                 Rapid to position #1
N30 G1 Z-.25                       Feed into workpiece
```

FIGURE 6.3 CIRCULAR INTERPOLATION USING THE CENTERPOINT METHOD

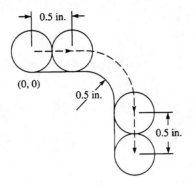

N40 X.5	Feed to position #2
N50 G2 X.5 Y-.5 I0 J.5	Cut CW arc to position #3
N60 G1 Y-1.	Feed to position #4
N70 G0 Z.1	Rapid to clearance
N80 X-4. Y0 M2	Rapid return to TC/PC End of program

Fig. 6.4 provides an example of cutting an arc in the Z-X plane using the G18 code.

The programming steps required to cut the arc shown in Fig. 6.4 using a 0.25 in. cutter are

N10 G0 G90 X-4. Y0 S2000 T1 M6	Rapid to TC/PC location
N20 X0 Z.1 F12.	Rapid to position #1
N30 G1 Z0	Feed into workpiece
N40 X-.625	Feed to position #2
N50 G18 G2 X.-1. Z-.5 I.5 K0	Cut CW arc to position #3
N60 G17 G1 X-1.5	Feed to position #4
N70 G0 Z.1	Rapid to clearance
N80 X-4. Y0 M2	Rapid return to TC/PC End of program

The Y-Z plane would be programmed similarly, replacing the G18 command with the G19 command and changing the programmed point to the necessary values of Y, Z, J, and K.

FIGURE 6.4　G18 PROGRAMMING

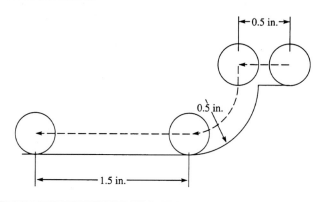

Another method of circular-interpolation programming involves using the X and Y words for the ending point and giving a radius for the arc using the R word rather than the I, J, and K words. Fig. 6.5 illustrates the use of X, Y, and R words for circular interpolation in cutting the arc in Fig. 6.4.

The programming steps required to cut the arc shown in Fig. 6.5 at a depth of 0.25 in. are

N10 G0 G90 X-4. Y0 S2000 T1 M6	Rapid to TC/PC
N20 X0 Z.1 F12.	Rapid to position #1
N30 G1 Z-.25	Feed into workpiece
N40 X.5	Feed to position #2

FIGURE 6.5　CIRCULAR INTERPOLATION USING THE RADIUS METHOD

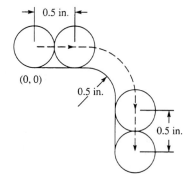

```
N50 G2 X1. Y-.5 R.5      Cut CW arc to position #3
N60 G1 Y-1.              Feed to position #4
N70 G0 Z.1               Rapid to clearance
N80 X-4. Y0 M2           Rapid return to TC/PC
                         End of program
```

Programming steps for a lathe and turning center are the same as given in the examples, but these machines use the following formats:

```
G2/G3 X. . . . Z. . . . I . . . . K. . . .
G2/G3 X. . . . Z. . . . R . . . .
```

6.4 Helical Interpolation

Helical interpolation is another important feature in CNC part programming. On two and a half-axis milling machines, it allows circular interpolation in two axes simultaneously (usually the X and Y axes), while providing linear-cut control in a third axis (usually the Z axis). Helical interpolation allows programming of helical pockets and threads.

Here is an example of cutting a helical thread on a machining center, a typical operation. It is as easy to cut a helical thread with a machining center as with a turning center, and is often quicker. This operation can be performed on a machining center without the extra setup time to put the workpiece on a faceplate or four-jaw chuck in a turning center. The helical thread can be cut in the same setup used to machine the other part features. In this example, only the thread will be cut. Circular interpolation in the X and Y axes and linear control in the Z axis will be used.

To program a helical thread, three things must first be determined: (1) the direction of the thread (right-hand or left-hand); (2) the number of turns of the thread (specified in the part drawing); and (3) the feed rate required to cut the thread.

Unless specified otherwise, all threads are right-hand threads. This requires a clockwise direction, where the thread advances in the clockwise direction. Left-hand threads advance in the counterclockwise direction. For example, when tightening a right-hand bolt, clockwise motion advances the bolt, while counterclockwise motion draws it out.

The number of thread turns and feed rate are obtained from the lead of the thread. All threads will be single lead unless otherwise specified in the part drawing. Single-lead threads mean one thread only.

The number of turns of the thread in the length of the thread determines the number of arcs to be cut. The linear feed rate for the Z axis (thread advance) depends on the depth the thread advances per turn. This information is determined from the lead of the thread. The lead of the thread is equal to the distance the thread advances per revolution. To determine the thread lead,

$$L = P \cdot I \qquad (6.1)$$

where L = lead of thread
P = thread pitch
I = number of leads on the thread

Thread pitch is equal to the reciprocal of the number of threads per inch. For example, the thread pitch of a 1/4-20 UNC, single-lead thread is 1/20 or 0.05. This means that the thread advances 0.05 in. per revolution. The lead and pitch of a single-lead thread will be the same. However, that does not mean that the two values are identical.

Dividing the overall length of the thread to be cut by the lead will determine the number of arcs required to cut the thread. For example, if we wanted to cut the 1/4-20 thread to a depth of 1 in., we would need 1.000/0.05 or 20 arcs. Not all threads work out to an even number of turns. Sometimes it is necessary to cut a fraction of a 360° arc at the end of the thread or to start the thread above the workpiece.

The feed rate for the Z axis is found by multiplying the spindle RPM by the lead of the thread.

$$F = \text{RPM} \cdot L \qquad (6.2)$$

If the spindle RPM is 250 and the lead of the thread is 0.05 in., the feed rate will be 12.5 IPM (inches per minute). A 60° thread milling cutter is used to cut threads on a machining center. Fig. 6.6 illustrates this setup.

The formula for calculating the depth of thread is

$$\text{single depth} = 0.6495/\text{number of threads per inch (TPI)} \qquad (6.3)$$

FIGURE 6.6 ☐☐☐☐☐☐☐☐☐☐☐☐☐☐☐☐☐☐☐☐☐☐☐☐☐☐☐☐
SETUP FOR MILLING THREADS

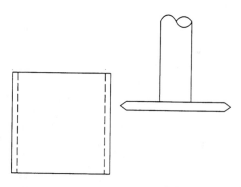

So, for 20 TPI we can determine a single depth of 0.6495/20 = 0.0325.

Suppose that we wanted to cut a 3 in.-diameter thread at 20 threads per inch. We want to make two passes, a roughing pass and a finish pass. We could use a subroutine with cutter diameter compensation, but for this example we will not use the subroutine. First, determine the coordinates of the thread depth by subtracting the thread depth from half the major diameter of the thread:

$$1.500 - 0.0325 = 1.4675$$

Then add the cutter radius. We will be using a 3 in.-diameter cutter, so

$$1.4675 + 1.5 = 2.9675$$

Then add the depth of the roughing pass to this coordinate. We will use a 0.005 in. depth for the finish pass, so

$$2.9675 + 0.005 = 2.9725$$

is our starting coordinate for the roughing pass.

Fig. 6.7 illustrates the programming steps necessary to cut the thread described. Not every CNC machine has helical interpolation capabilities. Check to determine whether your machines are equipped with helical interpolation.

Helical interpolation can be programmed in any one of the three plane pairs (X-Y, Y-Z, and Z-X) using the following word address formats:

```
G17 G14/G15 X. . . . Y. . . . I. . . . J. . . .
Z. . . . F. . . . L. . . .
G18 G14/G15 X. . . . Y. . . . I. . . . K. . . .
Z. . . . F. . . . L. . . .
G19 G14/G15 X. . . . Y. . . . J. . . . K. . . .
Z. . . . F. . . . L. . . .
```

FIGURE 6.7 ☐☐☐☐☐☐☐☐☐☐☐☐☐☐☐☐☐☐☐☐☐☐☐☐☐☐☐
HELICAL INTERPOLATION EXAMPLE

```
          Part Program for Milling Helical Thread
N100 G0 G90 X4. Y0 F10. S3000 T1 M6
N110 G0 X3. Z0
N120 G1 X2.9725
N130 G17 G14 X2.9725 Y0 Z-1. I0 J0 F12. L20
N140 X3.
N150 G0 Z0
N160 G1 X2.9675
N170 G17 G14 X2.9675 Y0 Z-1. I0 J0 F12. L20
N180 X3.
N190 G0 X4. Y0 M2
```

where G17, G18, and G19 select the appropriate plane pair. G14 and G15 specify the direction (G14 clockwise, G15 counterclockwise). X, Y, and Z coordinates are used to specify the arc endpoint coordinates. I, J, and K specify the arc centerpoint coordinates. F specifies the Z-axis feed rate. L is the number of complete 360° arcs to be cut.

6.5 SUMMARY

Linear interpolation is used to cut straight-line segments between two points. Linear interpolation can also be used to approximate arcs and curves; the more programmed segments, the better the approximation. The length and number of programmed segments depend on the maximum allowable tolerance of the arc.

Circular interpolation is used to cut arcs and circles. The G2 and G3 EIA standard G codes are used to initiate clockwise and counterclockwise cutter motion, respectively. The word address formats for these commands are as follows:

CENTERPOINT METHOD

```
G2/G3 X. . . .Y. . . .(Z. . . .)I. . . .J. . . .
(K. . . .)
```

where the X, Y, and Z words denote the ending point of the arc, and the I, J, and K words are used to program the centerpoint location of the arc.

RADIUS METHOD

```
G2/G3 X. . . .Y. . . .(Z. . . .)R. . . .
```

where the X and Y words are used to program the ending point of the arc, and the R word is used to program the radius of the arc.

Arcs may be cut in any of the plane pairs, X-Y, Z-X, or Y-Z. When programming in these alternate axes pairs, the appropriate G17, G18, or G19 code must be programmed first.

Helical interpolation is used to cut helical pockets and threads. Circular interpolation is used in two axes (X and Y typically), and linear control is used in a third (Z axis typically). Helical interpolation can be programmed in any one of the three plane pairs (X-Y, Y-Z, and Z-X) using the following word address formats:

```
G17 G14/G15 X. . . .Y. . . .I. . . .J. . . .
Z. . . .F. . . .L. . . .
G18 G14/G15 X. . . .Y. . . .I. . . .K. . . .
Z. . . .F. . . .L. . . .
```

```
G19 G14/G15 X. . . . Y. . . . J. . . . K. . . .
Z . . . . F . . . . L . . . .
```
where G17, G18, and G19 select the appropriate plane pair. G14 and G15 specify the direction (G14 clockwise, G15 counterclockwise). X, Y, and Z coordinates are used to specify the arc endpoint coordinates. I, J, and K specify the arc centerpoint coordinates. F specifies the Z-axis feed rate. L is the number of complete 360° arcs to be cut.

Before programming a helical thread, you need to know three things:

1. Direction of thread
2. Number of turns in the thread
3. Feed rate for cutting the thread

Not all machines have helical interpolation capabilities. Consult the machine's manual to determine the format and presence of helical interpolation programming.

Questions and Problems

1. What is the format for programming linear interpolation?
2. What two formats may be used for circular interpolation? Give examples of each.
3. What are the words I, J, and K and how are they used?
4. What is the word R used for?
5. Write the circular interpolation statements necessary to cut the first quadrant of a 3 in.-radius arc starting at the 3 o'clock position using the centerpoint method. The zero reference point is in the center of the arc.
6. Write a circular interpolation program for the arc in Problem #5 using the radius method.
7. What is helical interpolation?
8. How is helical interpolation used in CNC part programming?
9. What information must be known before programming a helix?
10. How is this information used to program the helix?
11. Program a 1.000-10 single-lead outside thread to a depth of 1 in. on a milling machine using helical interpolation in the X-Y plane.
12. Program a 2.000-10 single-lead outside thread to a depth of 2 in. using helical interpolation in the Z-X plane.

Programming CNC Lathes and Turning Centers

Chapter Objectives

After studying this chapter, the student will be able to

- Describe the setup information and procedures required prior to running a CNC program on a turning machine.
- Determine the necessary tooling information required in CNC programming.
- Recognize and program common operations performed on CNC turning machines.
- Recognize and program common canned cycles for CNC turning machines.

7.1 Introduction

This chapter will present information regarding CNC engine lathes, turret lathes, and turning centers. These machines turn the workpiece in a chuck and hold the cutter stationary, rather than turning the tool in a spindle while the workpiece remains in a fixed location, as in milling machines and machining centers. The examples given are generic by necessity. Specific controllers may use different commands and perform differently than those given in this text. Consult the machine manual before attempting to program for specific MCUs.

7.2 Setup Information

Fig. 7.1 depicts a CNC lathe similar to that discussed in this chapter. Basic CNC lathes have two axes, the X and Z axes. These are the primary axes on two-axis turning machines. Advanced CNC lathes and turning

FIGURE 7.1 CNC LATHE

(Courtesy of Hitachi Seiki USA)

centers may be equipped with four or more axes. These additional axes are generally designated U and W. Z-axis movement is parallel to the spindle or chuck axis. X-axis motion is the primary axis perpendicular to the spindle axis. A $(-)Z$ movement is toward the chuck, while a $(+)Z$ movement is away from the chuck. A $(-)X$ movement is toward the chuck centerline, while a $(+)X$ movement is away from the centerline. Programmable tailstocks, on lathes that are equipped with tailstocks, constitute one configuration of the third and fourth axes. Fig. 7.2 shows the basic axis arrangement for CNC lathes.

CNC lathes often have a *machine origin* based on a fixed machine coordinate system that becomes active as soon as the controller is turned on. These machines are fixed zero machines, discussed previously. The machine uses this coordinate system until a new local coordinate system is programmed. Typically, the fixed machine zero is located farthest to the right and away from the operator facing the machine on a back lathe. The programmer has the option of programming coordinates based on the fixed machine coordinates or establishing a local coordinate system. The machine coordinate system uses the U and W secondary motion-dimension words for programming. The local coordinate system uses X and Z motion-dimension words.

FIGURE 7.2
LATHE AXES

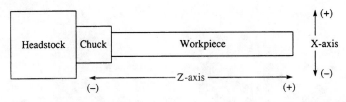

When you are setting up a CNC lathe, the first programmed movement should be a return to machine zero, using the G28 code. A G28U0W0 statement will move the machine back to the fixed zero location. The next programmed movement should bring the cutter a known distance away from the spindle centerline and chuck face. This is an absolute movement based on the machine zero reference point. After locating the cutter a known distance away from the desired X and Z axis points, program the distances the cutter is currently positioned away from these points using the G92, preset absolute registers command.

The G92 command specifies the distance the tool is currently located from the new zero location. The maximum chuck speed may also be programmed using the G92 code. Once the new local coordinate system has been established, coordinates may be programmed from the new zero reference. Machines that do not have the capability to return to machine zero and preset registers must be positioned manually. A note containing the necessary information should be appended to the setup sheet.

7.3 TOOLING INFORMATION

Either a fixed tool holder or a tool turret is used to hold the cutter on CNC lathes. Turning centers usually employ a tool turret capable of holding four or more tools at one time. These tools are typically of the carbide insert design, which are made to closer tolerances than conventional lathe tooling. When a tool-change statement is executed on the lathe, either the turret indexes to another tool position or an automatic tool change is executed, depending on the features available on the particular machine. Fig. 7.3 shows a CNC turning center.

The M6 function initiates a manual tool change. This function is programmed along with the tool number and tool offset register number using the **T address.** The format for the manual tool change is

$$M6 \; Tn_1n_2$$

FIGURE 7.3
CNC TURNING CENTER

(Courtesy of Hitachi Seiki USA)

where n_1 is the tool number

n_2 is the tool offset register number

When programming a turret tool change, the M6 function is not needed. Only the tool number and offset register number are programmed. The MCU will recognize the need for a tool change. The format for a turret tool change is

$$T n_1 n_2$$

where n_1 is the tool turret position

n_2 is the tool offset register number

The statement T201 would call tool position #2 and use the tool offset found in tool offset register #1.

It is often necessary to program a **dwell** to allow the tool turret time to position safely before further program execution begins. The G code used for programming a dwell is G4 P. . . , which specifies the amount of dwell in seconds. Dwells are used to allow time for the coolant to come on or for the chuck to come up to speed, and for turret rotation.

Each of the turning tools used on a CNC lathe has a tool nose radius. When programming lathe axis movements, the centerline or the edge of the tool nose radius (TNR) may be programmed; the choice is up to the

FIGURE 7.4 ☐☐☐☐☐☐☐☐☐☐☐☐☐☐☐☐☐☐☐☐☐☐☐☐☐☐
TOOL CENTERLINE VERSUS TOOL EDGE

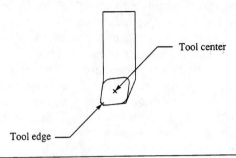

---

programmer. Fig. 7.4 illustrates the difference between tool centerline and tool edge.

One function of programmable tool offsets is to allow the programmer or operator to compensate for minor differences in length between different tools caused by tool replacement or tool wear. When the tools are set up, the appropriate tool offset is entered into the tool offset registers. When the tool number and offset combination is called, the controller then compensates by the offset value. This eliminates the need for preset tooling. Most tools and tool holder combinations are a preset length, for instance 4, 4.5, or 5 in. This allows the programmer to treat all tools as being the same length and not worry about calculating tool length offsets. Generally, the tool number will be the same as the tool offset register number. Cutter offsets cutter diameter compensation can still be programmed.

The following information must be determined and programmed prior to executing the part program:

Register number (positive number between 1 and the number of tool positions)

X-axis offset (signed number)

Z-axis offset (signed number)

Tool nose radius (positive value)

Standard tool nose vector number (positive number ranging from 1–12)

As stated previously, the tool may be programmed using the edge or the tool nose radius centerline. Tool-edge programming is adequate for straight-line cuts where part features intersect at right angles. However, problems arise when using edge programming to program angles and arcs. Using the tool edge will shift the centerpoint of the arc and induce an error in the programmed movement. The amount of error depends on the size of the tool nose radius and the arc radius. Tool centerline programming should be used when programming arcs and angles.

CNC lathes may be programmed using either radius values or diameter values for X-axis values, depending on the controller used. Some machines allow for either radius or diameter values, while other machines allow for only one. Diameter programming can be used for most turning and facing operations. Radius programming is used in threading operations. Many **canned cycles** require either radius or diameter values, depending on the particular cycle. Consult the manual to determine whether X-axis values should be programmed using radii or diameters. Programming the wrong values can lead to disastrous results.

Spindle speeds are specified by the **S address.** In lathes equipped with gear heads, spindle speeds are changed by shifting gears in the headstock. These machines usually provide two or more gear ranges. Therefore, a miscellaneous function is provided to change gear ranges. M40 through M46 are provided to change between speed ranges. For instance, M40 may be used for the low range, M41 for the next higher range, and so forth. Some machines are equipped with variable speed drives, which provide an infinite number of speeds within the low and high limits of the machine. The miscellaneous function is not needed on these machines.

In programming the CNC lathe, a G98 or G94 code is used to program feed rates in inches per minute. A G99 or G95 specifies feed rates in inches per revolution. For instance, G94/G98 F15. specifies a 15-inches-per-minute feed rate, while G95/G99 F.005 specifies a 0.005 inches-per-revolution feed rate.

In programming the CNC lathe, the coolant should be turned off during tool changes and back on during cutting. Coolants provide additional performance and protection for the tool. Adequate coolant

1. Reduces friction between the tool and workpiece.
2. Produces cleaner, more accurate cuts by allowing the tool to cut, rather than tear, the workpiece.
3. Washes out the chips produced in the operation.
4. Improves the surface finish and appearance of the finished part.
5. Prevents the buildup of material along the edges of the tool.
6. Extends the life of the tool.

Some general considerations concerning tooling are

1. Inspect all tools before they are used. Dull or broken tools should be replaced immediately. Check to make sure they are tight in the tool holder and the correct tool holder is being used.
2. Select the proper tool to perform the operation. Do not attempt to perform an operation for which the tool is not designed. Use the tool

properly and select the proper-sized tool to do the job safely and efficiently.
3. Maintain the proper cutting feeds and speeds for the tool used. Proper speeds and feeds maximize efficiency and tool life. Improper speeds and feeds waste time and money because of decreased tool life, slower production times, and tool changes.
4. Become acquainted with the machine's capabilities. Often, machines are used inefficiently because programmers do not know they can perform a simpler operation on the machine or are performing an operation incorrectly, causing excessive tool wear or tool breakage.
5. Program the largest depth of cut and highest feed rate safely possible. This reduces cycle time and has a negligible effect on tool life. Experience and common sense should guide your choice of depth of cut and feed rate. In situations where you are unsure, start small and build, based on feedback from the machine operator. Consult programming references such as machinist's handbooks or similar programs that cover the same operation and material.
6. Use coolants to extend tool life, clear chips, and reduce heat buildup in the workpiece.

7.4 Programming and Operating Procedures

The following program steps are made before running the part program:

1. Return to machine zero reference point.
2. Move incrementally to a safe tool change/part change location.
3. Program the zero point.
4. Identify and record the tool number, offset register number, X- and Z-axis offsets, tool nose radius, and tool nose vector numbers used in the program. Then enter these values into the controller's memory.
5. Record appropriate speeds and feed rates for each tool. The operator then has the option of using the override features of the machine's controller.
6. Load the part program into the controller's memory.
7. Dry-run the part program to check for error-free execution.
8. Load the workpiece into the machine, making sure that it is firmly held in position and that the zero point is located properly, as diagrammed on the setup sheet.

FIGURE 7.5
SETUP SHEET FOR A CNC LATHE OR TURNING CENTER

Lathe Setup Instructions

Part Identification							
Oper #	Tool #	Offset #	Position	X gage length	Z gage length		Remarks
X zero location Z zero location				Setup information and sketch			

9. Run the part program.
10. Fully inspect the first part produced by the program to determine the accuracy of the part.
11. If the part passes full inspection, continue production. If the part fails inspection, determine how and why the part failed. Send this information back to the programmer so that he or she can fix the problem. Fig. 7.5 is an example of a typical setup sheet for a CNC lathe. Fig. 7.6 is a flowchart of the steps used in programming CNC turning machines.

SECTION 7.4 PROGRAMMING AND OPERATING PROCEDURES 151

FIGURE 7.6
FLOWCHART OF PROGRAMMING STEPS

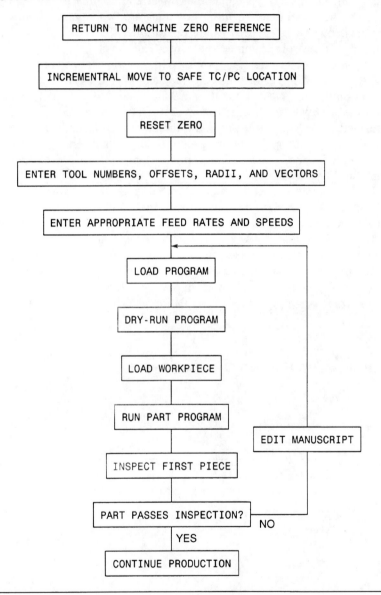

7.5 COMMON OPERATIONS PERFORMED

A CNC turning machine is capable of linear and circular interpolation in two axes simultaneously. Turning and facing are two common operations performed on the lathe. In turning operations, the rough stock is machined down to a programmed diameter. This involves moving the tool from an intermediate reference point (used for tool changes/part changes [TC/PC]) toward the chuck the desired length of the part feature. Depending on the depth of each cut, more than one pass is usually made to turn a piece down to a finished diameter. This turning operation can be performed through several program statements or through the use of a canned cycle. Roughing cuts are generally larger than finish cuts; therefore, roughing and finish cycles are available in lathe programming. Canned cycles will be discussed later in this chapter.

Facing operations are performed in the X-axis direction by bringing the tool to a programmed distance toward the chuck and then moving the tool toward or away from the spindle centerline. This allows the tool to cut a square shoulder on the workpiece. Turning and facing operations are often combined into one continuous cut, using two or more program statements. Turning and facing operations are illustrated in Fig. 7.7.

Taper turning is also an important feature in CNC lathe programming. Taper turning is simply linear interpolation used to make angular cuts on

FIGURE 7.7 ◻◻◻◻◻◻◻◻◻◻◻◻◻◻◻◻◻◻◻◻◻◻◻◻◻◻◻◻◻
TURNING AND FACING OPERATIONS

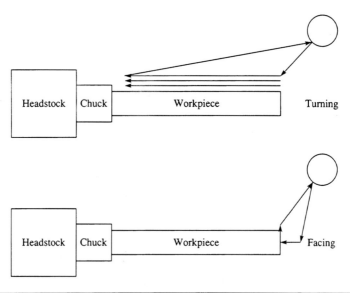

FIGURE 7.8 ☐☐☐☐☐☐☐☐☐☐☐☐☐☐☐☐☐☐☐☐☐☐☐☐☐
TAPER TURNING

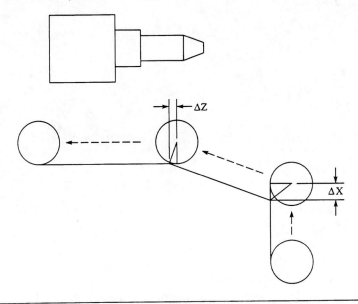

the lathe. The start and end points of the taper require cutter offsets. (The use of cutter offsets was discussed earlier.) One instance of using cutter offsets is to turn a taper such as the one illustrated in Fig. 7.8.

To save programming steps and time, operations may be performed using one or more of the following canned cycles, which are illustrated in Figs. 7.9 through 7.13:

FIGURE 7.9 ☐☐☐☐☐☐☐☐☐☐☐☐☐☐☐☐☐☐☐☐☐☐☐☐☐
SINGLE PASS THREADING CYCLE

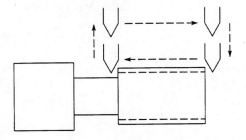

G33 SINGLE PASS THREADING CYCLE

The format for the G33 cycle is

$$G33\ Zn_1\ Kn_2$$

where n_1 is the length of the thread.

n_2 is the lead of the thread.

EXAMPLE

G33 Z1.K.05 cuts a thread with a pitch of 20 a length of 1 in.

G70 MULTIPLE REPETITIVE FINISHING CYCLE

This cycle must follow a G71 or G72 roughing cycle. The format for this cycle is

$$G70\ Pn_1\ Qn_2$$

where n_1 is the first cut statement in the finishing cycle.

n_2 is the last cut statement in the finishing cycle.

G71 MULTIPLE REPETITIVE ROUGHING CYCLE

When using the G71 cycle, first position the cutter at a clearance plane for the face and diameter of the workpiece. The X- and Z-axis motions must be in a generally increasing or decreasing direction. The format for this statement is

FIGURE 7.10 ◻◻◻◻◻◻◻◻◻◻◻◻◻◻◻◻◻◻◻◻◻◻◻◻◻◻◻
MULTIPLE REPETITIVE FINISHING CYCLE

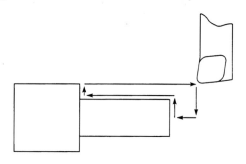

Figure 7.11
Multiple Repetitive Roughing Cycle

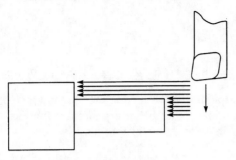

$$G71\ Pn_1\ Qn_2\ Un_3\ Wn_4\ Dn_5\ F\ S$$

where n_1 is the first cut statement. This statement must contain a single axis movement with a G0 or G1 programmed in the statement.

n_2 is the last cut statement.

n_3 is the amount of stock left on the diameter. This may be a positive or negative value, depending on whether the programmed movement is for a turning or boring operation.

n_4 is the amount of stock left on the face of the part. This value can also be positive or negative.

n_5 is the depth of each roughing cut. This is a radius value.

The roughing and finishing cycles are often used in conjunction, such as in the following example.

.	Previous cut statements
.	
.	
N200 G1 X1.1 Z2.1	Position at clearance in the X and Z axes
N210 G71 P220 Q260 U.005 W.005 D.05 F.008 S2500 N220 X0	Position at center of part
N230 Z2.	Position at finished length of part
N240 X.5	Cut to first diameter
N250 Z1.	Turn to length
N260 X1.	Cut to second diameter
N260 G0 X1.1 Z2.1	Return to clearance point

N270 G70 P210 Q250   Call finishing cycle
.   Additional part features
.
.

G72 Multiple Repetitive Rough Facing Cycle

The cutter is first positioned at a clearance for the face and the diameter. X- and Z-axis motions must be in a generally increasing or decreasing direction. The format for the G72 statement is

$$G72 \; Pn_1 \; Qn_2 \; Un_3 \; Wn_4 \; Dn_5 \; F \; S$$

where n_1 is the first cut statement. This statement must contain a single axis movement with a G0 or G1 programmed in the statement.

n_2 is the last cut statement.

n_3 is the amount of stock left on the diameter. This may be a positive or negative value, depending on whether the programmed movement is for a turning or boring operation.

n_4 is the amount of stock left on the face of the part. This value can also be positive or negative.

n_5 is the depth of each roughing cut. This is a radius value.

G76 Multiple Pass Threading Cycle

The format for the multiple pass threading cycle is

$$G76 \; Xn_1 \; Zn_2 \; In_3 \; Kn_4 \; Dn_5 \; Fn_6 \; An_7$$

Figure 7.12 □□□□□□□□□□□□□□□□□□□□□□□□□□□
Multiple Repetitive Rough Facing Cycle

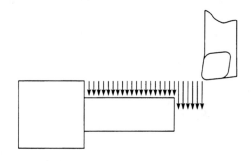

**FIGURE 7.13
MULTIPLE PASS THREADING CYCLE**

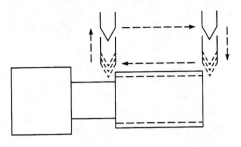

where n_1 is the minor diameter of the thread.

n_2 is the length of the thread.

n_3 is the difference in the thread radius from one end of the thread to the other. This word address is used for cutting tapered threads. For straight threads, a value of zero is entered.

n_4 is the height of the thread from the crest to the root. This is a radius value.

n_5 is the lead of the thread.

n_6 is the feed rate.

n_7 is the angle of the tool tip. This is usually 60 degrees.

EXAMPLE

G76 X.9693 Z1.10 K.0307 D.05 F10. A60.

Modern turning machines are capable of machining threads. These threads may be constant lead straight, tapered, or multiple-start. To produce a thread on a turning machine, the cutter is first positioned to depth at the proper starting distance from the workpiece. The proper cycle code is given and the thread is cut. After cutting the thread, the tool retracts and returns to the starting position, ready for the next threading pass. This process is repeated until the final thread depth is reached. The thread is produced by synchronizing the cutter feed rate in IPR with the spindle speed in RPM. The tool advancement per spindle revolution is equal to the lead of the thread previously discussed.

The following are miscellaneous functions used specifically during lathe programming.

M3 Spindle ON with CW Rotation (normal rotation for a front lathe, i.e., programmed movements are made while looking at the front of the chuck.

This is contrasted with a rear lathe in which the opposite is true.)
M4   Spindle ON with CCW Rotation
M5   Spindle and Coolant Stop
M8   Coolant ON
M9   Coolant OFF
M10 Clamp Chuck
M11 Unclamp Chuck
M13 Spindle ON CW Rotation and Coolant ON
M14 Spindle ON CCW Rotation and Coolant ON
M30 End of Program
M80–92 Indexes the Chuck from positions 1-12
(M80 releases the chuck.)

7.6 EXAMPLES

EXAMPLE 7.6 (A) ☐☐☐☐☐☐☐☐☐☐☐☐☐☐☐☐☐☐☐☐☐☐☐☐☐
TURNING/FACING/CIRCULAR INTERPOLATION

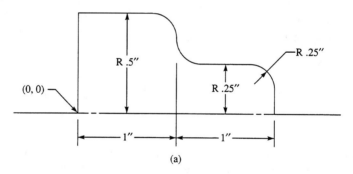

(a)

```
N10  G28 U0 W0                            (Return to Machine Zero)
N20  G0 U-2 W-1                           (Incremental movement)
N30  G92 X5 Z4                            (Preset Absolute Registers)
N40  G95 G96 F.005 S1750 M13
N50  G0 X0
N60  Z2.1
N70  G71 P80 Q150 U.005 W.005 D.03 F.005 S1750   (Roughing Cycle)
N80  G1 Z2
N90  X.25
N100 G3 X.5 Z1.75 I0 K.25
N110 G1 Z1.25
N120 G2 X.75 Z1 I.25 K0
N130 G3 X1 Z.75 I0 K.25
N140 G1 Z0
N150 X1.125
N160 G70 P80 Q150 F.002 S2500             (Finishing Cycle)
N170 G0 X5 Z4 M30                         (End of Program)
```

EXAMPLE 7.6 (B) ☐☐☐☐☐☐☐☐☐☐☐☐☐☐☐☐☐☐☐☐☐☐☐☐☐
TURNING/FACING/THREADING INTERPOLATION

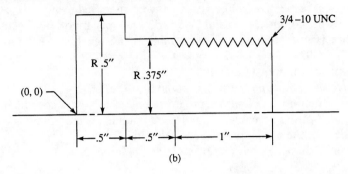

(b)

```
N10  G28 U0 W0                        (Return to Zero Reference)
N20  G0 U-2 W-1                       (Incremental movement)
N30  G92 X5 Z4 S3500                  (Preset Absolute Registers)
N40  G95 G96 F.005 S1750
N50  T101 M13
N60  G0 X0
N70  Z2.1
N80  G71 P90 Q140 U.005 W.005 D.03    (Roughing Cycle)
N90  G1 Z2
N100 X.75
N110 Z.5
N120 X1
N130 Z0
N140 X1.125
N150 G70 P90 Q140 F.002 S2500         (Finishing Cycle)
N160 G0 X5 Z4 T202
N170 X.725 Z2.1
N180 G33 Z1 K.1 F.1                   (Threading Cycle)
N190 G0 X.825
N200 Z2.1
N210 G1 X.7
N220 G33 Z1 K.1 F.1
N230 G0 X.8
N240 Z2.1
N250 G1 X.685
N260 G33 Z1 K.1 F.1
N270 G0 X5
N280 Z4 M30                           (End of Program)
```

7.7 SUMMARY

There are two primary axes on a CNC lathe or turning center—the X and Z axes. If the machine is equipped with more than two axes, these additional axes are designated U and W. Rigid tool holders or tool turrets are used to hold tools on CNC turning equipment. Tool offsets must be entered into the controller before executing a part program. Standard tool nose vector numbers are used to describe the tool orientation when using tool nose radius compensation.

Tool numbers and offset register numbers are programmed using the T address. Spindle speeds and tool feeds are programmed using the S and F addresses, respectively. Feed rates may be specified using the G94 code for inches-per-minute feeds and G95 for inches-per-revolution feeds. The G28 code is used to return the machine to the machine zero reference point. The G92 code is used to preset an intermediate zero point for successive programming statements.

There are two types of programming on the lathe—diameter and radius programming. In diameter programming, the X-axis coordinates are actually half of the actual tool movement. Radius programming is used to move the tool the actual programmed distance.

Several canned cycles are available in lathe programming. These cycles are

G33 Single pass threading cycle
G70 Multiple repetitive finishing cycle
G71 Multiple repetitive roughing cycle
G72 Multiple repetitive rough facing cycle
G76 Multiple pass threading cycle

The following miscellaneous functions are available for lathe programming.

M3   Spindle ON with CW rotation
M4   Spindle ON with CCW rotation
M5   Spindle and coolant stop
M8   Coolant ON
M9   Coolant OFF
M10 Clamp chuck
M11 Unclamp chuck
M13 Spindle ON CW rotation and coolant ON
M14 Spindle ON CCW rotation and coolant ON
M30 End of program
M80–92 Indexes the chuck from positions 1–12.

QUESTIONS AND PROBLEMS

1. What are the two primary axes of motion for turning machines?
2. What is the difference between radius and diameter programming? How is each normally used?
3. What is the code and format of the single pass threading cycle?

**FIGURE 7.14
PROBLEM #11**

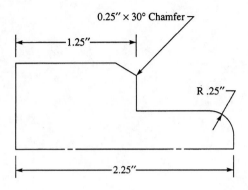

4. What is the code and format for the multiple repetitive finishing cycle?
5. What is the code and format for the multiple repetitive roughing cycle?
6. What is the code and format for the multiple repetitive rough facing cycle?
7. What is the code and format for the multiple pass threading cycle?
8. Write a program using the multiple repetitive roughing and finishing cycles to turn and face a 1.5 in.-dia. piece of raw stock down to a 1 in. dia. that protrudes 3.100 in. from the chuck. The finished length of the piece should be 2.5 in., leaving 0.5 in. from the chuck as clearance.
9. Write the statement using the single pass threading cycle necessary to cut a 1/4-20 UNC thread 1 in. long.
10. Write the statement using the multiple pass threading cycle necessary to cut a 1/2-13 UNC thread 1 in. long.
11. Write the program statements necessary to turn and face the part shown in Fig. 7.14 using the centerpoint and radius methods of circular interpolation and absolute positioning.

Programming CNC Milling Machines and Machining Centers

Chapter Objectives

After studying this chapter, the student will be able to

- Describe the setup information and procedures required prior to running a CNC program on a milling machine.
- Determine the necessary tooling information for a milling machine.
- Recognize and program common operations performed on CNC milling machines.
- Recognize and program common canned cycles for CNC milling machines.

8.1 Introduction

This chapter will provide concepts and procedures associated with programming numerically controlled milling machines and machining centers. CNC milling machines and machining centers perform a number of different operations including milling, drilling, boring, counterboring, reaming, and tapping. Machining centers have had the most significant impact on the manufacturing industry of all CNC machines. These machines are able to perform all the previously mentioned operations in a single workpiece setup. Additionally, machining centers are capable of changing their own tools with automatic tool changers. It is due to these additional capabilities that the CNC machining center has gained such an important role in the modern manufacturing industry.

These machines range from two- or three-axis machines, which utilize point-to-point positioning and manual tool changers, up to five-axis contouring machining centers, which have automatic tool-change capabilities. This chapter will describe the capabilities of these machines, setup information required for programming, tooling, and common operations performed on these machines. Pertinent program examples will also be provided.

8.2 Setup Information

After loading the program to be executed into the controller, one of the first setup procedures involves setting the zero point or part origin. This is the point where the X, Y, and Z axes intersect, the basis for all subsequent absolute movements. The programmer should specify this point on the setup sheet.

The part origin is set by jogging the machine's table until the spindle is directly over the point at which the part origin will be located. Then, special dials, zero-set buttons on the controller, or special G codes tell the MCU that this is the new zero location.

After establishing a part origin, the machine is moved to the tool change/part change location. This is a point off to one side of the work table. It allows for removal of the workpiece and tool holders as new tools or parts are required. The programmer specifies the TC/PC location on the setup sheet as one of the first lines in the program.

After the TC/PC location has been established and the machine has been located there, the workpiece to be machined may be loaded and clamped in the workholding fixture. The workpiece may be clamped directly to the table, clamped in a vise, or mounted on a fixture.

8.3 Tooling Information

Now the operator is ready to enter tooling data. This data may be entered by the programmer, if he or she knows the specific parameters such as tool length, offset, etc. that will be used. Typically, these values are entered by the operator. One of the first operations is determining the preset length for each tool: measuring and recording the length and calculating the necessary tool length offset for each tool used. The necessary tools should be specified on the setup sheet in the order in which they will be used. The operator then manually enters the tool number, tool length offset, and tool diameter, giving the machine's controller access to this information during execution of the program.

In addition to the number, offset, and diameter of each tool, the spindle speeds and feed rates programmed for each tool should be entered on the setup sheet. These figures are supplied by the programmer and entered into the program and are given on the setup sheet for the operator's information. Should it be necessary, and the machine so equipped, the operator may choose to use the speed and/or feed rate override features of the machine. Many controllers have a program check page that appears on the graphics display. This program check page shows the programmed feed rate and spindle speed as well as sequence number, current programmed block, tool number, and other relevant programmed information.

Modern machining equipment is often equipped with precision surface-sensing equipment, or "probe". This probe is used to electronically program the X, Y, or Z reference point and provide feedback for automatic positioning compensation. It is brought into physical contact with the workpiece, where the MCU knows how far the probe extends from the machine. Using this as a reference, the MCU can determine a position in space based on this reference. The probe can be used for a number of operations such as locating the setup point, establishing TLOs, or locating off of part features. The probe is generally located in the tool magazine, where it is loaded automatically. Fig. 8.1 illustrates the precision surface-sensing tool or probe.

Once the previous steps have been completed, the operator should dry-run a new program to make sure that the program accomplishes what it is supposed to accomplish. This means that the program will execute without interference with machine travel limits, fixed machine features, fixtures, etc. and that it is free of errors. During dry-run operations, programmed statements may be executed block-by-block or in a continuous mode.

The operator may choose to dry-run using the rapid override, in which programmed movements at the rapid feed rate are reduced to allow the operator to stop program execution before any damage can occur. Another dry-run choice is to increase programmed movements

FIGURE 8.1
PRECISION SURFACE-SENSING TOOL

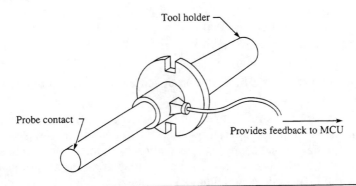

at feed rates to the rapid rate. This reduces the dry-run execution time; however, it also reduces the time the operator has to react to errors. The part program may be dry-run while cutting using the block-by-block mode or operation. While in this mode, a single programmed statement is executed and the MCU waits for a key stroke to continue. This allows for the first part to be produced while testing for proper feed rate, spindle speeds, and other programmed parameters. Be aware that there are a number of options available to the operator during dry-run procedures.

Once the operator is satisfied with the program's execution, the program is run for the first time. After the workpiece has been machined, a full inspection of the piece is made to determine its accuracy. If the part passes inspection, production continues. If the part does not meet specifications, the operator should discontinue the production run and note how the program failed. The program is then sent back to the programmer for revision.

The following steps are made prior to and during the actual running of a part program:

1. Locating and resetting the part origin.
2. Establishing the tool change/part change location (unless specified by the tool builder or tool-change mechanism) and moving there.
3. Measuring and recording the tool number, tool length offsets, and tool diameters used in the program. These values are then entered into the controller's memory.
4. Recording appropriate speeds and feed rates for each tool. The operator then has the option of using the override features of the machine's controller.
5. Loading the part program into the controller's memory.
6. Dry-running the part program to check for error-free execution.
7. Loading the workpiece in the machine, making sure that it is firmly clamped in position and that the part origin is located properly, as diagrammed on the setup sheet.
8. Running the part program.
9. Fully inspecting the first part produced by the program to determine its accuracy.
10. If the part passes full inspection, continuing production. If the part fails inspection, the operator should try to determine how and why the part failed and send this information back to the programmer to correct the problem. Fig. 8.2 is an example of a typical setup sheet for milling machines and machining centers.

FIGURE 8.2
MILL SETUP SHEET

Mill Setup Instructions

Part Identification								
Oper #	Tool #	Description	Length	TLO	Tool descr.	RPM	Feed	Notes

X zero location
Y zero location
Z zero location

Setup information and sketch

8.4 COMMON OPERATIONS PERFORMED

Point-to-point milling machines are used in operations such as drilling, counterboring, and countersinking, where the machine performs an operation at a single point in space. These machines offer limited capabilities and are manufactured for specific operations.

Linear-cut machines allow straight-line or linear cuts parallel to one axis at a time. Operations performed include face and pocket milling. Two- and three-axis linear-cut machines allow linear cuts simultaneously in more than one axis. However, they do not allow circular interpolation. One, two, or three axes may be controlled simultaneously for straight-line cuts at any angle. These linear-cut machines are among the lowest-cost

CNC machines available, due to the reduced cost of the MCU. However, their low cost is offset by their limited capabilities.

CNC contouring machines are the most popular numerically controlled machines today. CNC contouring machines provide the additional capability of circular interpolation in two or three axes simultaneously. Some machines offer two-axis simultaneous control and linear interpolation in a third axis. These machines are called two and a half-axis machines and are capable of point-to-point positioning and linear-cut control, in addition to machining curves in space with a minimum of program statements. Fig. 8.3 illustrates a CNC milling machine.

The CNC milling machines that offer the greatest versatility are machining centers. Capable of up to five-axis simultaneous control, they can machine all sides of the workpiece in a single setup. Some machining centers are also capable of combining milling/drilling operations with turning operations because they can provide both tool spindle rotation and workpiece rotation. Machining centers are capable of normal X, Y, and Z (longitudinal, transverse, and vertical) axes motions and vertical and horizontal swivel motions. In addition, machining centers are often equipped with automatic tool changers. This feature distinguishes machining centers from milling machines. Fig. 8.4 shows a typical CNC machining center.

In addition to linear and circular interpolation, the following canned cycles (illustrated in Fig. 8.5 through Fig. 8.12) are available on most CNC milling machines and machining centers:

G77 FACE MILLING CYCLE

This allows the programmer to face the surface of a part with an end mill or face mill. The format for this cycle is

$$G77\ Xn_1\ Yn_2\ Yn_3\ F$$

where n_1 is the distance to be traveled in the X direction.

n_2 is the distance to be traveled in the Y direction.

n_3 is the amount of stepover for each pass.

Note: This process is faster than the pocket milling cycle. The cutter position should be programmed so that at least the radius of the cutter extends beyond the part on all sides.

EXAMPLE

G77 X6.5 Y6.5 Y.5 F20 would face-mill a 5 in. × 5 in. workpiece with a 0.5 stepover in the +Y direction between passes using a 3 in.-diameter face mill.

FIGURE 8.3
CNC MILLING MACHINES

(Courtesy of Bridgeport Machines, Inc.)

Figure 8.4
CNC Machining Center

(Courtesy of Cincinnati Milacron)

FIGURE 8.5
FACE MILLING CYCLE

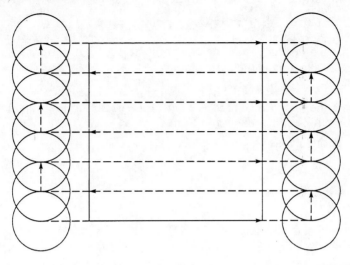

FIGURE 8.6
POCKET MILLING CYCLE

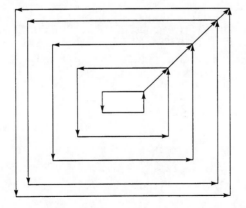

G78 Pocket Milling Cycle

The format for the pocket milling cycle is

$$G78\ Xn_1\ Xn_2\ Yn_3\ Yn_4\ F$$

where n_1 is the distance from the center of the pocket to the wall along the X axis minus the cutter radius.

n_2 is the X-axis stepover value.

n_3 is the distance from the center of the pocket to the wall along the Y axis minus the cutter radius.

n_4 is the Y-axis stepover value.

EXAMPLE

G78 X3. X.25 Y3. Y.25 F10 would mill a 3 in. × 3 in.-pocket with a stepover between each pass of (.25,.25) using a 0.25 in.-end mill.

G79 Internal Hole Milling Cycle

This cycle is used to bring a hole to the proper size using a milling cutter. The format for this cycle is

Figure 8.7 ☐☐☐☐☐☐☐☐☐☐☐☐☐☐☐☐☐☐☐☐☐☐☐☐☐☐☐☐
Internal Hole Milling Cycle

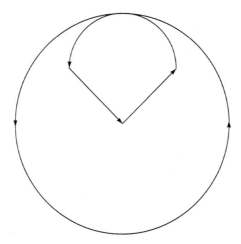

$$G79\ Jn\ F$$

where *n* is the radius of the hole to be milled minus the cutter radius.

EXAMPLE

G79 J1. F12. would mill a 2.5 in.-diameter hole using a 0.5 in.-diameter cutter.

G81 DRILLING CYCLE

This cycle provides a feed in, rapid out sequence suitable for drilling a series of holes of the same diameter and depth. G81 is modal, remaining in effect until a G0, G80, or other fixed cycle is programmed. The G80 code is used to cancel the operation of canned cycles. Z*n* is the total unsigned incremental distance that the tool will travel. The format for the drilling cycle is

$$G81\ Xn_1\ Yn_2\ Zn_3\ Zn_4\ F$$

where n_1 is the X-axis coordinate location of the hole to be drilled.

n_2 is the Y-axis coordinate location of the hole to be drilled.

n_3 is the total unsigned incremental distance of tool travel.

n_4 is the reference plane the cutter will rapid-out to after drilling the hole.

**FIGURE 8.8
DRILLING CYCLE**

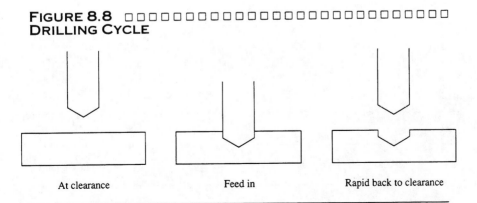

EXAMPLE

G81 X2. Y1. Z1.1 Z.1 F10 would drill a hole 1 in. deep from a clearance plane of Z.1.

G82 Drilling Cycle with Dwell

This cycle performs the same function as the drilling cycle, but allows a dwell at the bottom of the hole. Zn is the total unsigned incremental distance that the tool will travel. The format for this cycle is

$$G82\ Xn_1\ Yn_2\ Zn_3\ Zn_4\ Fn_5\ Pn_6$$

where n_1 is the X-axis coordinate location of the hole to be drilled.

n_2 is the Y-axis coordinate location of the hole to be drilled.

n_3 is the total unsigned incremental distance of tool travel.

n_4 is the reference plane.

n_5 is the feed rate of the drilling operation.

n_6 is the amount of dwell time in seconds that the drill will remain at the bottom of the hole.

EXAMPLE

G82 X2. Y1. Z1.1 Z.1 F10. P2 would drill a 1 in.-deep hole from a Z.1 clearance plane and remain at the bottom of the hole for 2 seconds.

FIGURE 8.9
Drilling Cycle with Dwell

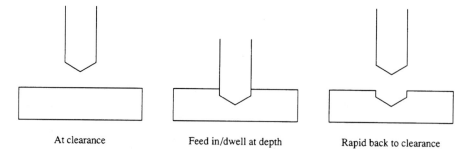

FIGURE 8.10
DEEP HOLE DRILLING CYCLE

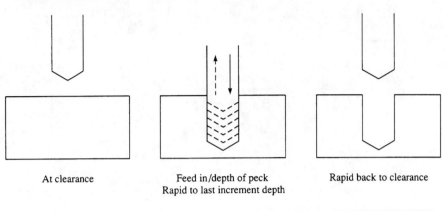

At clearance   Feed in/depth of peck   Rapid back to clearance
Rapid to last increment depth

G83 DEEP HOLE DRILLING CYCLE

This cycle provides a feed in, rapid out, rapid in to the bottom of the hole, feed in, rapid out to reference, rapid in to the bottom of the hole, feed in, and so on until the final hole depth is reached. G83 is also modal. The format for this cycle is

$$G83\ Xn_1\ Yn_2\ Zn_3\ Zn_4\ Zn_5\ F$$

where n_1 is the X-axis coordinate location of the hole to be drilled.

n_2 is the Y-axis coordinate location of the hole to be drilled.

n_3 is the total unsigned incremental distance of tool travel.

n_4 is the incremental distance for the first peck increment.

n_5 is the peck increment for successive passes. n_4 is the default if n_5 is not programmed.

EXAMPLE

G83 X2. Y1. Z1.1 Z.25 Z.1 F10 would peck-drill a 1 in.-deep hole from a clearance plane of Z.1 with a first increment of Z − .25 and successive increments of Z − .1.

FIGURE 8.11 TAPPING CYCLE

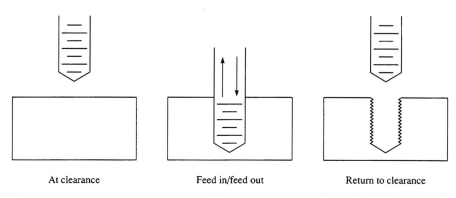

At clearance   Feed in/feed out   Return to clearance

G84 TAPPING CYCLE

This cycle provides a feed in, feed out sequence appropriate for tapping operations using a tapping attachment for nonreversing spindles. G84 is also modal. The format for this cycle is

$$G84 \ Xn_1 \ Yn_2 \ Zn_3 \ Zn_4 \ F$$

where n_1 is the X-axis coordinate location of the hole to be drilled.

n_2 is the Y-axis coordinate location of the hole to be drilled.

n_3 is the total unsigned incremental distance of tool travel.

n_4 is the reference plane the cutter will rapid-out to after drilling the hole.

EXAMPLE

G84 X2. Y1. Z1.1 Z.1 F7 would tap a hole 1 in. deep from a clearance plane of Z.1.

G85 BORING CYCLE

This cycle provides a feed in, feed out sequence appropriate for boring or reaming operations. The G85 cycle is modal. This cycle is programmed the same as the drilling cycle:

$$G85 \ Xn_1 \ Yn_2 \ Zn_3 \ Zn_4 \ F$$

where n_1 is the X-axis coordinate location of the hole to be drilled.

FIGURE 8.12 BORING CYCLE

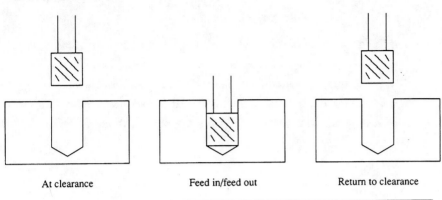

At clearance   Feed in/feed out   Return to clearance

n_2 is the Y-axis coordinate location of the hole to be drilled.

n_3 is the total unsigned incremental distance of tool travel.

n_4 is the reference plane the cutter will rapid-out to after drilling the hole.

G86 BORING CYCLE

This cycle provides a feed in, feed stop, wait for operator command, rapid out, operator restart sequence suitable for boring operations. The G86 cycle is modal. This cycle is programmed the same as the drilling cycle, allowing the operator control over the position of the cutter during exit. The G86 cycle program format is

$$G86 \; Xn_1 \; Yn_2 \; Zn_3 \; Zn_4 \; F$$

where n_1 is the X-axis coordinate location of the hole to be drilled.

n_2 is the Y-axis coordinate location of the hole to be drilled.

n_3 is the total unsigned incremental distance of tool travel.

n_4 is the reference plane the cutter will rapid-out to after drilling the hole.

G87 CHIP-BREAKING CYCLE

This cycle is programmed and operates the same as the deep hole drilling cycle except that instead of rapiding-out of the hole and back after the feed move, the G87 cycle will rapid-up and return a distance of 0.050 in.

(1.3 mm). This breaks the chip rather than withdrawing the tool from the work each peck as in the G83 cycle. The format for the G87 cycle is

$$G87 \ Xn_1 \ Yn_2 \ Zn_3 \ Zn_4 \ Zn_5 \ F$$

where n_1 is the X-axis coordinate location of the hole to be drilled.

n_2 is the Y-axis coordinate location of the hole to be drilled.

n_3 is the total unsigned incremental distance of tool travel.

n_4 is the incremental distance for the first peck increment.

n_5 is the peck increment for successive passes. n_4 is the default if n_5 is not programmed.

G89 BORING CYCLE

This cycle provides a feed in, dwell, feed out sequence appropriate for boring operations. The dwell causes the surface perpendicular to the spindle axis to be machined with no surface cutter defects. The G89 cycle is modal. The format for this cycle is

$$G89 \ Xn_1 \ Yn_2 \ Zn_3 \ Zn_4 \ F$$

where n_1 is the X-axis coordinate location of the hole to be drilled.

n_2 is the Y-axis coordinate location of the hole to be drilled.

n_3 is the total unsigned incremental distance of tool travel.

n_4 is the reference plane the cutter will rapid-out to after drilling the hole.

Note: The dwell time in seconds is set by a previously programmed G4 P*n* statement.

Several miscellaneous functions are used during the programming of milling machines and machining centers: M0, M1, M2, M6, and M25.

M0 PROGRAM STOP

An M0 stops the programming cycle and allows the operator to perform some function such as tool transfer. Instructions outlining the function to be performed during the program stop should be given on the setup sheet. The program stop takes effect after the machine motion in the block in which it is programmed has been executed.

M1 Optional Stop

This function performs the same as the M0 program stop except that the operator must switch the **OPSTOP** function on the controller to the ON position. If the OPSTOP function is not turned on, the machine will ignore the M1 function. The function will take effect after the programmed machine movement in the block in which it was programmed has been executed.

M2 End of Program

The code M2 executes a program stop and retracts the spindle to the home position before executing the last machine movement. It also resets the sequence number to the first program line and enables the operator to continue the production run after changing tools and/or parts.

M6 Tool Change

Tool-change requests are made using the M6 code. When this code is programmed, the MCU retracts the spindle to the home position and stops the spindle. The operator changes the tool and restarts the spindle. The next block is then executed.

M25 Home Axis

This function retracts the spindle to the Home position at the rapid traverse rate. If X and Y movements have been programmed, they are executed after the spindle has retracted.

8.5 Examples

Example 8.5 (A) ☐☐☐☐☐☐☐☐☐☐☐☐☐☐☐☐☐☐☐☐☐☐☐☐☐
Circular Interpolation and Drilling Cycle

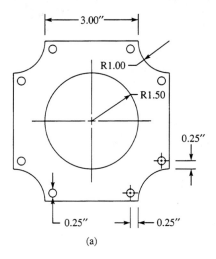

(a)

```
N10  G0 G90 X-5 Y0 F20 S2000 T1 M6    (1" End Mill Cutter)
N20  X-3 Y-2
N30  Z-.25
N40  G2 X-2 Y-2.5 R.5
N50  G1 Y-3
N60  X2
N70  G2 X2.5 Y-2 R.5
N80  G1 X3
N90  Y2
N100 G2 X2 Y2.5 R.5
N110 G1 Y3
N120 X-2
N130 G2 X-2.5 Y2 R.5
N140 G1 X-3
N150 Y-2
N160 Z.1
N170 G0 X0 Y0
N180 G1 Z-.5
N190 G79 J1 F8                        (Internal Hole Mill Cycle)
N200 Z.1
N210 G0 X-5 Y0 F16 S1800 T2 M6        (.25" Drill)
N220 G81 X-1 Y-2 Z1.1 Z.1             (Drilling Cycle)
N230 X1 Y-2
N240 X2 Y-1
N250 X2 Y1
N260 X1 Y2
N270 X-1 Y2
N280 X-2 Y1
N290 X-2 Y-1
N300 G80
N310 G0 X-5 Y0 M2                     (End of Program)
```

Example 8.5 (b)
Linear and Circular Interpolation

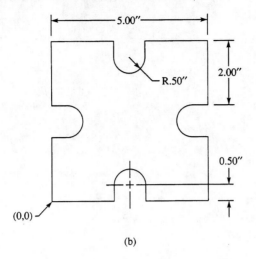

(b)

```
N10  G0 G90 X-5 Y0 F12 S1200 T1 M6        (.5" End Mill Cutter)
N20  X0 Y-.25
N30  G1 Z-.375
N40  X2.25
N50  Y.5
N60  G2 X2.75 Y.5 R.25
N70  G1 Y-.25
N80  X5.25
N90  Y2.25
N100 X4.5
N110 G2 X4.5 Y2.75 R.25
N120 G1 X5.25
N130 Y5.25
N140 X2.75
N150 Y4.5
N160 G2 X2.25 Y4.5 R.25
N170 G1 Y5.25
N180 X-.25
N190 Y2.75
N200 X.5
N210 G2 X.5 Y2.25 R.25
N220 G1 X-.25
N230 Y-.25
N240 G0 Z.1
N250 X-5 Y0 M2                            (End of Program)
```

Example 8.5 (c)
Drilling/Tapping/Face Mill/Pocket Cycles

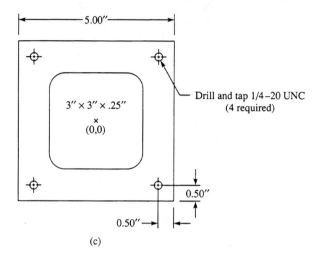

(c)

```
N10  G0 G90 X-5 Y0 F20 S2000 T1 M6      (3" Face Mill Cutter)
N20  X-2.5 Y-2.5
N30  G1 Z-.1
N40  G77 X5 Y5 Y1 F8                    (Face Mill Cycle)
N50  G0 Z.1
N60  X-5 Y0 F10 S1200 T2 M6             (1" End Mill Cutter)
N70  G0 X0 Y0
N80  G1 Z-.35
N90  G78 X2.5 X.375 Y2.5 Y.5 F10        (Pocket Milling Cycle)
N100 G0 Z.1
N110 X-5 Y0 F18 S1800 T3 M6             (3/16" Drill)
N120 Z.1
N130 G81 X-2 Y2 Z1.5 Z.1                (Drilling Cycle)
N140 X2 Y2
N150 X2 Y-2
N160 X-2 Y-2
N170 G80
N180 G0 X-5 Y0 F6 S100 T4 M6            (1/4-20 UNC Tap)
N190 Z.1
N200 G84 X-2 Y2 Z1.5 Z.1                (Tapping Cycle)
N210 X2 Y2
N220 X2 Y-2
N230 X-2 Y-2
N240 G80
N250 G0 X-5 Y0 M2                       (End of Program)
```

8.6 Summary

CNC milling machines and machining centers range from two- and three-axis point-to-point machines through five-axis contouring machines. These machines are set up in basically the same way. The following steps outline the procedures used in setting up the machine prior to, and steps performed during, the running of a part program:

1. Locating and resetting the part origin.
2. Establishing the tool change/part change location (unless specified by the tool builder or tool change mechanism) and moving there.
3. Measuring and recording the tool number, tool length offsets, and tool diameters used in the program. These values are then entered into the controller's memory.
4. Recording appropriate speeds and feed rates for each tool. The operator then has the option of using the override features of the machine's controller.
5. Loading the part program into the controller's memory.
6. Dry-running the part program to check for error-free execution.
7. Loading the workpiece in the machine, making sure that it is firmly clamped in position and that the part origin is located properly, as diagrammed on the setup sheet.
8. Running the part program.
9. Fully inspecting the first part produced by the program to determine its accuracy.
10. If the part passes full inspection, production commences. If the part fails inspection, it is necessary to determine how and why the part failed and relay this information back to the programmer.

Milling machines and machining centers may be used to drill, tap, ream, bore, counterbore/countersink, and mill or perform a combination of these operations. Many canned cycles are available which simplify programming. These cycles include

G77 Face Milling Cycle
G78 Pocket Milling Cycle
G79 Internal Hole Milling Cycle
G80 Canned Cycle Cancel
G81 Drilling Cycle
G82 Drilling Cycle with Dwell
G83 Deep Hole Drilling Cycle
G84 Tapping Cycle

G85 Boring Cycle
G86 Boring Cycle
G87 Chip-Breaking Cycle
G89 Boring Cycle

The following miscellaneous functions are used when programming the milling machine or machining center:

M0   Program Stop
M1   Optional Stop
M2   End of Program
M6   Tool Change
M25 Home Axis

Questions and Problems

1. List the steps performed prior to starting a production run.
2. What distinguishes a CNC milling machine from a CNC machining center?
3. What types of operations are performed on CNC machining centers?
4. What are the G code, format, and function for the drilling cycle?
5. Write the program statement that would cut a 1 in. deep hole at coordinates (3,5) and a clearance of Z.1 using the drilling cycle.
6. What are the G code, format, and function for the deep hole drilling cycle?
7. Write the program statement necessary to cut a 3-in. deep hole at coordinates (4.5,7.25) at a clearance of Z.1 using the deep hole drilling cycle.
8. What are the G code, format, and function for the tapping cycle?
9. Write the program statement that would tap a 0.75-in. deep thread at coordinates (1.75,4) and a clearance of Z.1 using the tapping cycle.
10. What are the G codes, formats, and function for the boring cycles?
11. Write the program statement necessary to bore a 2.5-in. deep bore at coordinates (5,7) and a clearance of Z.1 using the G86 boring cycle.
12. What are the G code, format, and function for the chip-breaking cycle?

FIGURE 8.13 □□□□□□□□□□□□□□□□□□□□□□□□□
PROBLEM #14

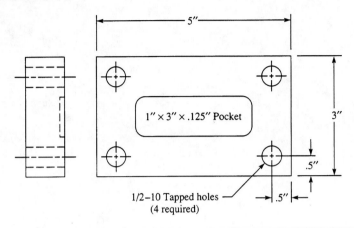

13. Write the program statement needed to cut a 2.75-in. deep hole at coordinates (1,1) with a clearance of Z.1 and an increment of 0.25 for all pecks using the chip-breaking cycle.
14. Write the program statements necessary to face-mill, mill the pocket, drill, and tap the part shown in Fig. 8.13.
15. Write the program statements necessary to mill the part shown in Fig. 8.14 using the centerpoint and radius methods of circular interpolation and absolute positioning.

FIGURE 8.14 □□□□□□□□□□□□□□□□□□□□□□□□□
PROBLEM #15

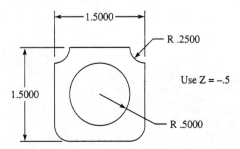

OTHER MACHINES PROGRAMMED USING NUMERICAL CONTROL

CHAPTER OBJECTIVES

After studying this chapter, the student will be able to
- Recognize and describe common operations performed on CNC punching machines.
- Define the concept of nesting.
- Recognize and describe the operations performed on CNC electrical discharge machines.
- Recognize and describe the operations performed on CNC flame-cutting machines.

9.1 INTRODUCTION

Greater emphasis has been placed on the use of numerical control in machining operations such as milling and turning. While these operations are very important and make up most numerical control operations, CNC technology has been applied to a growing range of applications in metalworking and nonmetalworking operations.

New applications of numerical control technology are constantly being explored and developed. This chapter will investigate some of the more common applications, such as CNC punching machines, electrical discharge machining (EDM), and flame cutting. CNC applications are limited only by the imagination of the engineer and the economics of the application. Almost anything that can be broken down into finite steps can be programmed using numerical control technology.

9.2 Numerical Control Punching Machines

CNC punches, known by a variety of terms, including punch presses, fabricators, and turret punches, do not have a rotating spindle or chuck. They are used for punching holes in cabinets, panels, and patterns. Manual punch presses are still used for specialized work, but numerical control provides a 300 to 1500% increase in the amount of work throughput, at a substantially reduced scrap and rework rate.

These machines are available in a variety of sizes and capacities. Many options and accessories are available, such as tapping attachments for threading holes and flame cutters for cutting larger and irregularly shaped cutouts. These machines are all numerically controlled. Most punching machines fall in the 30- to 40-ton (267 to 356 kN) capacity range, while some have capacities of 100 tons (890 kN) or greater. These larger machines are able to pierce up to 1.125 in. (28 mm) mild steel plate! Fig. 9.1 shows typical CNC punching machines.

Most punching machines are made to run at a rate of 60 to 120 strokes per minute, at a maximum rate of 200 strokes per minute for short bursts. The faster the stroke rate, the greater wear and tear on tooling and machine parts. Stroke rate should be programmed to minimize process time, while maximizing tool life, machine life, and safety.

Programming CNC punching machines is relatively simple, yet, due to the large number of strokes needed to punch larger holes and cutouts, programs often take a great deal of time to develop manually. Thus, the use of the computer reduces programming time and maximizes the number of pieces that may be punched from a sheet of stock. This latter feature is called *nesting*.

FIGURE 9.1 ◻◻◻◻◻◻◻◻◻◻◻◻◻◻◻◻◻◻◻◻◻◻◻◻◻◻◻◻
CNC PUNCHING MACHINES

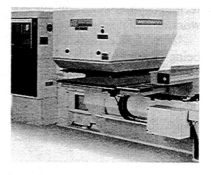

SECTION 9.2 NUMERICAL CONTROL PUNCHING MACHINES   189

Nesting is often done by a separate program that takes each part feature separately and moves them into different positions on the piece of stock. The programmer enters the stock size and programmed part features. The nesting program then optimizes part placement to reduce scrap and increase machine efficiency. If only a few parts are required from a single piece of stock, nesting may be done manually. However, when programming several parts from a single piece of stock, the computer is often faster and more efficient. Fig. 9.2 illustrates the nesting concept.

Each CNC punching machine manufacturer uses a different coding system. However, most companies stick fairly close to the EIA standard. Before attempting to program any CNC machine, check the operator's manual to determine the specific coding system used by the machine's MCU.

Fig. 9.3 shows selected examples of work done by CNC punching machines. Included are

1. Punching holes: round, square, hexagonal, and special shapes typi-

FIGURE 9.2
NESTING

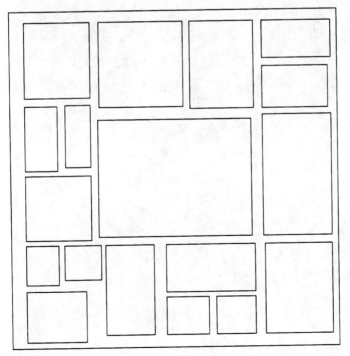

FIGURE 9.3
CNC PUNCHING MACHINE EXAMPLES

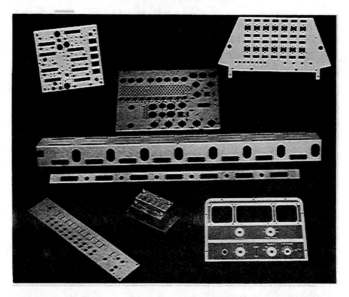

(Courtesy of Murata Wiedemann, Inc.)

cally ranging from 0.050 to 5 in. (1 to 125 mm) in diameter. The shape of the hole depends on the shape of the punch used to make the hole.

2. Notching typically involves taking square pieces off the corners or edges, although notches may not necessarily be square.
3. Nibbling is accomplished through a series of short equally spaced steps that produce curved or diagonal cutouts. Nibbling leaves a scalloped edge that can be smoothed for a better finish.
4. Knockouts are holes that are pierced except for one or more small tabs that hold the knockout in place until removal is required. These knockouts are used in applications such as electrical service boxes, wall boxes, and electrical cabinets.
5. Louvers are used for ventilation. They are often used in air conditioning, ventilator covers, and applications that direct the flow of air over or through panel covers.
6. Other shapes and features available on punching machines are countersinking, extruding, cornering, and odd shapes such as keyholes that require special punches. Larger features can be nibbled, while smaller shapes are better produced by single strokes.

Punches and dies come in a variety of shapes and sizes: square, round, hexagonal, etc. in sizes from 0.050 to 5 in. (1 to 125 mm) in diameter. These punch and die sets have different clearances depending on the material and stock thickness used. Periodically, these sets become dull and must be replaced or sharpened.

Punch and die sets are placed in the machine along with a punch holder. Punching machines may be equipped with only one tool position or may have multiple-station capability. Multiple-station punching machines use tool turrets similar to the automatic tool changers described earlier.

Single-punch machines have greater tonnage capacities and are made to handle larger sizes of stock, but usually perform only one or a few operations. Single-punch machines require that the machine be moved to a tool-change position that allows the punch and die set to be removed. The current set is replaced by the new set and the machine repositioned. Thus, like manual tool changers on CNC milling machines, tool-change times are greater for single-punch machines than for turret-type punching machines. One example of a single-punch machine is shown in Fig. 9.4.

Turret-type machines are available with 10 to 38 tool stations. These machines often offer several heavy-duty tool positions for thicker materials and larger diameter holes. The remaining tool positions are light- to medium-duty. Turret-type machines reduce the tool-change times by carrying the necessary tools within the turret. The operator must set up and install the necessary tools prior to running the program. These punch and

FIGURE 9.4 □□□□□□□□□□□□□□□□□□□□□□□□□□□
SINGLE-PUNCH CNC MACHINE

(Courtesy of Trumpf America, Inc.)

die sets are specified on the setup sheet supplied by the programmer. Fig. 9.5 shows a turret-type CNC punching machine.

Stock thicknesses generally run from 18 gage (0.048 in. [1.2 mm]) to 0.25 in. (6 mm). Stock sizes vary according to throat depth and work table sizes. General sizes are 30 × 36 in. (760 × 910 mm) to 48 × 60 in. (1210 × 1520 mm). The majority of punched parts are made from mild steel, which has a shear strength of 50,000 psi (345 MPa). Other common materials include stainless steel, galvanized steel, brass, and plastic. The maximum hole size that can be punched in these materials depends on the shear strength of the material and the material's thickness. The thicker the material or the higher the shear strength, the smaller the maximum hole that can be punched in the material for a given machine capacity.

Workpieces are generally held on the work table by pneumatically operated clamps. Small air cylinders clamp and unclamp the workpiece to the table. Once the workpiece is clamped to the table, the table moves in the X and Y directions throughout the program. Stops or "bumps" are often used to locate the workpiece so that subsequent workpieces may be

FIGURE 9.5
TURRET-TYPE CNC PUNCHING MACHINE

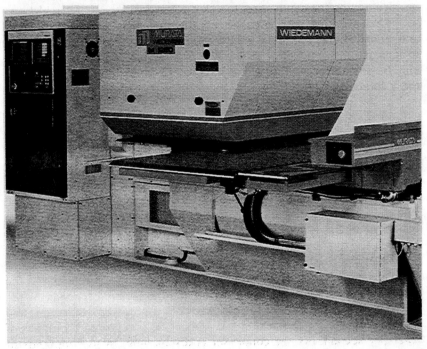

(Courtesy of Murata Wiedemann, Inc.)

TABLE 9.1
SPECIFIC CODES USED FOR CNC PUNCHING MACHINES

Code	Explanation
G1	Linear nibbling mode
G2	Circular nibbling mode, CW
G3	Circular nibbling mode, CCW
G67	Punch off
G68	Punch on
G69	Tool retract, punch off, return to home position
G90	Absolute programming
G92	Preset register
G94	Incremental programming
M6	Calls tool 1 (start-up position)
M12	Cycle stop, optional stop
M71	Tape rewind
M75	Punch load position
T	Random tool selection

accurately located. These clamps are often movable so that conflicts between clamps and part features can be avoided. Clamp locations should be specified on the setup sheet to avoid collisions.

When working with larger sheets the workpiece can be unclamped and moved. Longer sheets can be moved to the right or left as necessary, to increase the available travel in the X direction. Also, the workpiece may be unclamped and turned around to effectively double the travel in the Y direction.

9.3 ELECTRICAL DISCHARGE MACHINING

Electrical discharge machining (EDM) removes material through high-energy electric sparks that flow from an electrode to the workpiece. Fig. 9.6 shows an EDM machine. The tool and workpiece are submerged in a fluid bath with a low electrical conductivity, such as light oil. A CNC servo system maintains a very small gap, approximately 0.001 in. (0.025 mm), between the tool and the workpiece while cutting.

When the voltage across the gap between the tool and workpiece reaches a sufficient level, condensers discharge a high current across the gap, which causes a spark, for a period of 10 to 30 microseconds. When the voltage drops to around 12 volts, the condensers stop discharging and start to recharge for the next spark. This cycle is repeated thousands of times per second. As the condenser is discharging, many thousands of sparks are produced between the tool and the workpiece around the shape of the tool. Each of these sparks removes a small amount of material. This process is similar to the buildup and release of energy in a car's distributor by means of the ignition points, although the process is controlled so that the points do not burn or pit. Fig. 9.7 shows a schematic of the EDM process.

In addition to the fixed-electrode EDM process, the tool electrode may be a soft copper wire about 0.008 in. (0.20 mm) in diameter. This wire electrode is slowly fed from a supply spool to a take-up spool to compensate for the erosion of the wire and particles that accumulate on the wire. In wire-EDM, the dielectric (low electrical conductivity fluid surrounding the gap between the tool and the workpiece) is usually deionized water. Fig. 9.8 shows a wire-EDM machine.

Wire-EDM allows the manufacturing of intricate shapes and openings, small-radius contours (internal and external), and the cutting of thin stock materials without specially shaped tooling. Numerical control technology is used to provide up to four-axis positioning, which maintains the proper gap between the tool and workpiece while providing accurate positioning. Fig. 9.9 illustrates the wire-EDM process.

The EDM process may be used for any good electrical conducting material, including metals, their alloys, and most carbides. Properties

FIGURE 9.6
ELECTRICAL DISCHARGE MACHINE

(Courtesy of AGIE)

FIGURE 9.7
SCHEMATIC OF THE EDM PROCESS

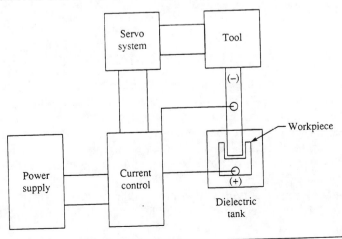

FIGURE 9.8
WIRE-EDM MACHINE

(Courtesy of Elox Division, Colt Industries)

such as melting point, hardness, toughness, and brittleness of the stock do not limit the use of the EDM process. EDM can be used to form holes or shapes in materials that may be too hard or brittle to be machined economically by other methods. Since there is no direct physical contact between the tool and the workpiece, very delicate work may be done and very thin materials may be machined. Fig. 9.10 shows some pieces made by the EDM process.

9.4 FLAME-CUTTING CNC MACHINES

CNC flame-cutters are generally used to cut shapes out of large, thick plates. Typically, many parts are cut from a sheet of stock. Functions controlled numerically include marking and scribing for part layout, piercing, preheating, acetylene and oxygen supply control for oxyacetylene cutting, torch height, template tracing, and plasma arc cutting. With the addition of any one of a variety of nesting programs available, parts can be produced accurately and efficiently.

Fig. 9.11 shows a typical thermal machining center. There are two types of CNC flame-cutters: oxyacetylene and plasma arc. Most CNC

FIGURE 9.9
THE WIRE-EDM PROCESS

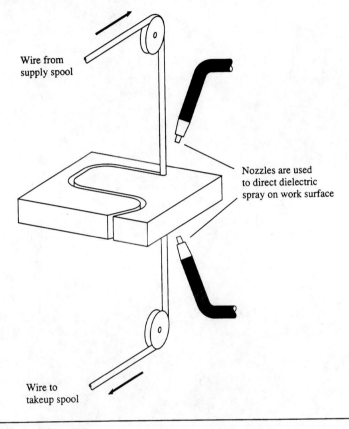

flame cutters are gantry-type, where a rigid, box-shaped bridge equipped with axis drive motors controls torch positioning. Cutting speeds depend on the thickness of the material, generally ranging up to 240 in./min or more. CNC flame cutters are available in a wide variety of sizes up to 16 ft wide and 40 or more feet long.

Flame-cutters may be equipped with one or more cutting heads. Larger machines hold up to 10 cutting heads at one time. Flame-cutters can perform the following operations: contouring, linear cuts, squaring, edge preparation, beveling, and curve fitting. They are used to produce parts for pressure vessels, engine frames, large equipment parts, ship plate, and other applications.

CNC flame-cutters are generally programmed using EIA standard G and M codes. When programming a CNC flame-cutter, however, become

FIGURE 9.10 ☐☐☐☐☐☐☐☐☐☐☐☐☐☐☐☐☐☐☐☐☐☐☐☐☐☐☐
PARTS MANUFACTURED BY THE EDM PROCESS

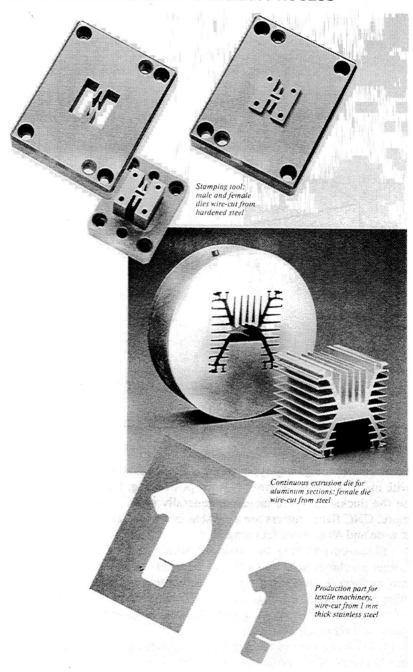

Stamping tool: male and female dies wire-cut from hardened steel

Continuous extrusion die for aluminum sections: female die wire-cut from steel

Production part for textile machinery, wire-cut from 1 mm thick stainless steel

(Courtesy of AGIE)

Figure 9.11
Thermal Machining Center

(Courtesy of MG Industries)

acquainted with the specific MCU's programming format before beginning to program the machine. No standard coding format is required of manufacturers.

In CNC flame-cutting operations, the workpiece is lowered into the flame-cutting machine's water tank to prevent heat buildup and subsequent warpage. The workpiece is generally positioned using a hoist or crane for larger pieces, although a forklift may be used on smaller pieces. The workpiece is lowered onto a worktable and may be clamped. Sometimes the workpiece is left free, since the cutter does not contact the workpiece. The workpiece is raised or lowered so that the water in the tank just covers it. If the worktable is stationary, the water level may be raised or lowered as needed.

9.5 Summary

We have covered just some of the many other applications of numerical control technology beside milling and turning operations. CNC punching machines, electrical discharge machining (EDM), and flame-cutting machines all add an additional range of flexibility in machining through the use of numerical control.

Questions and Problems

1. How does CNC programming for punching machines differ from milling and turning?
2. What is nesting and how is it accomplished?
3. List five major functions performed by CNC punching machines.
4. Describe single-punch and turret-type punching machines.
5. Describe the EDM process.
6. How is the wire used in the wire-EDM process?
7. What are the advantages of the EDM process over other methods?
8. What are the two major types of flame-cutting?
9. List five major functions that may be performed on CNC flame-cutting machines.
10. What is the function of the water tank in CNC flame-cutting?

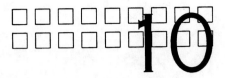

Repetitive Programming and Advanced Features

Chapter Objectives

After studying this chapter, the student will be able to
- Recognize and describe the concept of looping as it relates to CNC programming.
- Recognize and describe the use of subroutines within CNC programs.
- Describe and apply the concept of mirror imaging to CNC programming.

10.1 Introduction

Repetitive programming allows the programmer to define a set of instructions to be used more than once within the same program. This reduces the program development time and the number of required program statements. Repetitive programming involves the use of loops and subroutines. Nesting involves placing a loop within another loop or a subroutine or placing a subroutine within another subroutine, all of which provide additional flexibility in programming. Also included in this chapter are the concepts of threading and tapping, which are accomplished through the use of specific G codes. Additional calculations concerning speeds, feed rates, and thread calculations are necessary for programming threads and using taps.

10.2 LOOPING

Looping allows the programmer to jump back to an earlier part of the program and execute the programmed movements a specified number of times. It reduces the program length and redundancy of writing these instructions more than once. One format for looping is

$$= Na/b$$

where a is the loop end block sequence number.

b is the number of times the loop is to be repeated.

The range for the loop is from the programmed statement following the looping statement down to and including the loop end block statement.

Once the controller encounters a looping statement, the program statements within the loop are executed. When the controller executes the final statement in the loop, the controller decrements a register holding the number of times the loop is to be executed. The controller then jumps control back to the first statement after the looping statement and executes the loop again. This process continues until the register holding the number of times the loop is to be executed contains zero. When the controller reaches the end block statement and the repeat register contains zero, the controller continues with the next statement following the loop.

EXAMPLE

$$= N100/3$$
```
N100 G91 G81 X1 Z.5 F200
```

This will cause the machine to drill three holes beginning 1 in. away from the current position and indexing 1 in. in the positive X direction each time the loop is executed. This is due to the incremental positioning mode, G91. The drill cycle, G81, was used to further reduce the number of programming steps.

Fig. 10.1 gives an example of the use of looping.

Another method of looping involves the G codes, G51 and G50. The format for this method is

```
G51 Nn
G91 G81 X1 Z.5 F200
G50
```

This will have the same result that the previous loop did. G51 is the looping code. N*n* tells the controller how many times the loop is to be executed. The loop starts on the next programmed statement. The range runs from the statement following the G51 code to the G50 code, which designates the loop end.

FIGURE 10.1 LOOPING

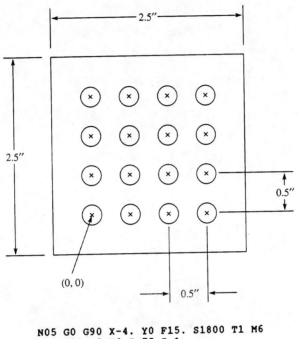

```
N05 G0 G90 X-4. Y0 F15. S1800 T1 M6
N10 G81 X0 Y0 Z.75 Z.1
=N15/3
N15 G91 X.5
N20 Y.5
=N25/3
N25 X-.5
N30 Y.5
=N35/3
N35 X.5
N40 Y.5
=N45/3
N45 X-.5
N50 G0 G90 X-4. Y0 M2
```

Nesting loops involves placing one loop within another. The number or levels of nested loops depends on the particular controller's capabilities. The same format and procedures are used to nest loops. Particular attention should be given to nesting; it is easy to make errors in incremental positioning within nested programming. Fig 10.2 is an example of nested loops.

Some restrictions to keep in mind with looping, depending on the controller, include

1. The loop end block sequence number must appear later in the program.

FIGURE 10.2
NESTED LOOPING

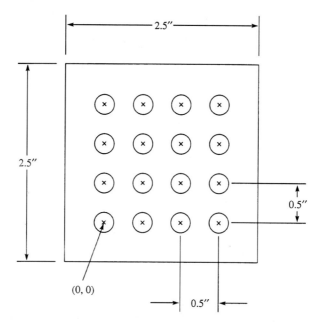

```
N05 G0 G90 X-4. Y0 F18. S2500 T1 M6
=N30/2
N10 G91 G81 X0 Y0 Z.75 Z.1
=N15/3
N15 X.5
N20 Y.5
=N25/3
N25 X-.5
N30 Y.5
N35 G0 G90 X-4. Y0 M2
```

2. The block following the looping statement must have a sequence number.
3. Loops may be nested. The number of nested loops varies with the controller used. However, the range of nested loops must lie within the range of the next outer loop. They can share the same end block sequence number.
4. The maximum number of repetitions for a loop depends on the capabilities of the controller. All controllers have some practical limit on the maximum number of times a loop may be repeated, depending on memory and register capacity.
5. The loop end block cannot be the last statement in the program. This

is typically averted by programming a movement to the part/tool change position following loop execution.
6. Errors can be introduced through using incremental positioning with looping. Pay particular attention to zero shifts and cycle programming when using loops and nesting loops.

10.3 SUBROUTINES

Subroutines or **macros** are small programs contained within larger programs. They allow for changes in parameters when calling the subroutine. A number assigned to the subroutine is used to call the subroutine, which is usually placed at the beginning of the program. When the program is executed, the controller reads the subroutine and stores it in memory until it is called for later in the program. One format for subroutine definitions is

The #n statement defines the subroutine as the nth subroutine. The subroutine ranges from the next program statement to the subroutine end symbol ($). Within subroutines, variable parameters are assigned the symbol (*) following the appropriate word.

The statement used to jump to the macro within the main program is known as the *call statement*. The format for the subroutine call statement is

$$= \#n \ F*100$$

where n is the subroutine to be executed and the variable parameter F is assigned the value of 100.

The number of macros that may be defined within a single program is dependent on the specific controller used.

EXAMPLE

```
#1
N1 G92 X0 Y0
N2 G90 G81 X1. Y0 Z1.1 Z.1 F*
N3 Y1.
N4 X0
N5 G80
$
```

This is defined as subroutine number 1. Statement N1 shifts the program zero reference to the current tool location. Statement N2 tells the controller that absolute positioning is to be used and that it should drill a 1-in. deep hole from a clearance plane of Z.1 at the (1,0) location. Statement N3 tells the controller to move to the (1,1) location. Statement N4 tells the controller to move to the (0,1) location. Statement N5 tells the controller to cancel the canned cycle. The symbol ($) terminates the subroutine and control is returned to the main program. Fig. 10.3 gives an example of a subroutine and call statements used in CNC programming.

A subroutine may also be called using the miscellaneous functions, M98 and M99. These function numbers will vary depending on the controller used. In this example, M98 calls the subroutine and M99 terminates the subroutine:

N150 Pn_1 Ln_2 M98 n_1 specifies the subroutine number. n_2 is the number of times to execute the subroutine.

n_1
|
Subroutine
statements
|
M99   Returns control to main program.

Some restrictions to keep in mind concerning subroutines include

1. One subroutine is not defined within another subroutine. However, one subroutine may be called from within another subroutine.
2. Variables are assigned in the order they are defined. The number of variable parameters in the subroutine and the number of assigned values in the call statement must be the same.
3. Subroutines may include loops. A call statement for another subroutine may be included in the loop within the macro. The subroutine termination symbol ($) must not be the end block statement of the loop.
4. Subroutines may be nested several levels, depending on the capabilities of the controller.
5. As many variables as necessary may be included in the subroutine. However, the call statement may be restricted to a certain number of characters. This is the limiting factor when assigning variables to the subroutine.

Nesting also involves defining loops within loops or defining subroutines within other subroutines. The concept of nesting is demonstrated in the following program, where a loop is defined within a loop and both are contained within a subroutine. This short program drills the series of holes shown in Fig. 10.4.

FIGURE 10.3
SUBROUTINE EXAMPLE

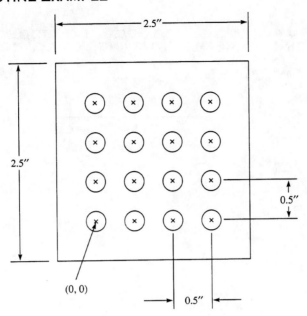

```
#1
N100 G91 G81 X0 Y0 Z.75 Z.1 F18.
=N105/3
N105 X.5
$
N05 G0 G90 X-4. Y0 S1800 T1 M6
N10 X0 Y0
=#1
N10 G0 G90 X0 Y.5
=#1
N15 G0 G90 X0 Y1.
=#1
N20 G0 G90 X0 Y1.5
=#1
N25 G0 G90 X-4. Y0 M2
```

To introduce the concept of mirror programming, the same program could be written as:

```
#1
N100 G* G91 G81 X0 Y0 Z.75 Z.1 F18.
=N105/3
N105 X.5
$
N05 G0 G90 X-4. Y0 S1800 T1 M6
N10 X0 Y0
=#1 G*30
N10 G0 G90 Y.5
=#1 G*31
N15 G0 G90 Y1.
=#1 G*30
N20 G0 G90 Y1.5
=#1 G*31
N25 G0 G90 X-4. Y0 M2
```

FIGURE 10.4 □□□□□□□□□□□□□□□□□□□□□□□□□□□□
SUBROUTINE PROGRAM CONTAINING NESTED LOOPS

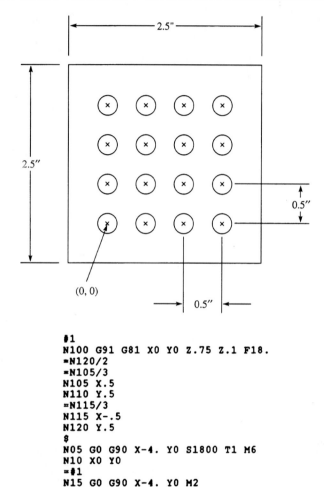

```
#1
N100 G91 G81 X0 Y0 Z.75 Z.1 F18.
=N120/2
=N105/3
N105 X.5
N110 Y.5
=N115/3
N115 X-.5
N120 Y.5
$
N05 G0 G90 X-4. Y0 S1800 T1 M6
N10 X0 Y0
=#1
N15 G0 G90 X-4. Y0 M2
```

10.4 MIRRORING

Mirroring reverses the sign of programmed axis movements. It can be used to cut a symmetrical pattern in all four quadrants or continue cutting a symmetrical feature in all four quadrants. An example of mirror programming is given in Fig. 10.5.

FIGURE 10.5
Mirroring Example

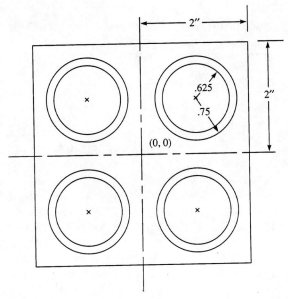

```
#1
N100 G*
N105 X1.6875 Y1.
N110 Z-.25
N115 G3 X1. Y1.6875 I.6875 J0
N120 X.3125 Y1. I0 J.6875
N125 X1. Y.3125 I.6875 J0
N130 X1.6875 Y1. I0 J.6875
N135 X0 Y0 Z.1
$
N05 G0 G90 X-4. Y0 F20. S1800 T1 M6
N10 X0 Y0 Z.1
=#1 G*30
=#1 G*32
=#1 G*31
N15 G30
=#1 G*31
N20 G30 G0 G90 X-4. Y0 M2
```

Programming the arc segments using the radius method

```
#1
N100 G*
N105 X1.6875 Y1.
N110 Z-.25
N115 G3 X1. Y1.6875 R.6875
N120 X.3125 Y1. R.6875
N125 X1. Y.3125 R.6875
N130 X1.6875 Y1. R.6875
N135 X0 Y0 Z.1
$
N05 G0 G90 X-4. Y0 F20. S1800 T1 M6
N10 X0 Y0 Z.1
=#1 G*30
=#1 G*32
=#1 G*31
N15 G30
=#1 G*31
N20 G30 G0 G90 X-4. Y0 M2
```

Recall from Chapter 3 that the values in the four quadrants are

Quadrant	X	Y
1st Quadrant	positive	positive
2nd Quadrant	negative	positive
3rd Quadrant	negative	negative
4th Quadrant	positive	negative

In order to reproduce the features of a part programmed in the first quadrant, the X values must be changed from positive to negative. The part origin may be located in the center of the workpiece for convenience so that mirrored part features are centered on the workpiece. The G code used to change the X values from positive to negative is G31. The Y values may be similarly reversed by using the code G32. The G31 and G32 codes are modal, remaining in effect until canceled by another code. In this case, G30 is programmed to cancel any programmed mirroring. Mirroring is also called *symmetry programming* or *mirror imaging*.

Frequently, a subroutine is written containing the first-quadrant features of the programmed part. A variable parameter may be used to substitute the required G code for the second, third, and fourth quadrants. For example,

= #1 G*30 would cut the first quadrant.

= #1 G*31 would cut the second quadrant.

= #1 G*32 (G31 is already in effect) would cut the third quadrant.

= #1 G*32 (where a G30 command had been previously programmed outside of the subroutine to cancel all mirroring) would cut the fourth quadrant.

An example of mirroring is given in Fig. 10.6.

```
#1
N100G*G91G0X.5Y.5
N105G1Z-.225
N110X2.
N115Y2.
N120X-2.
N125Y-2.
N130Z.225
N135G0X-.5Y-.5
$
N5G90G0X-4.Y0T01M6
N10X0Y0Z.1S2500
=#1G*30
=#1G*31
=#1G*32
N15G30
=#1G*32
N20G0G90X-4.Y0
N25M2
```

FIGURE 10.6
MIRRORING

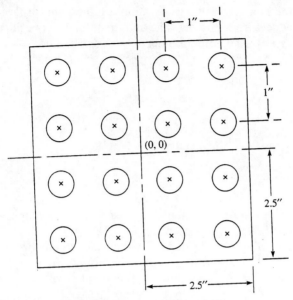

```
#1
N100 G*
N105 X1.6875 Y1.
N110 Z-.25
N115 G3 X1. Y1.6875 I.6875 J0
N120 X.3125 Y1. I0 J.6875
N125 X1. Y.3125 I.6875 J0
N130 X1.6875 Y1. I0 J.6875
N135 X0 Y0 Z.1
$
N05 G0 G90 X-4. Y0 S1800 T1 M6
N10 X0 Y0 Z.1
=#1 G*30
=#1 G*32
=#1 G*31
N15 G30
=#1 G*31
N20 G30 G0 G90 X-4. Y0 M2
```

10.5 SUMMARY

A loop allows the controller to execute a series of programmed statements a number of times, which reduces the program length and programming time. The format for looping is

$$= Na/b$$

where a is the loop end block sequence number and b is the number of times that the loop will be executed.

A subroutine is a small program within a large main program. Subroutines also reduce programming time and the number of required statements. They allow the programmer to assign values to variable parameters defined within the subroutine. The format for subroutines is

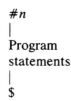

To call the subroutine, the statement = #n is given, along with any assigned variables, such as = #n F*200.

Nesting allows loops to be defined within loops or subroutines and subroutines to be defined within other subroutines. It provides added flexibility in programming and reduces the number of program statements.

Mirroring, mirror imaging, or symmetry programming is used to cut features programmed in the first quadrant in other quadrants. The codes G30, G31, and G32 are used to change the sign of programmed axis moves. Mirroring is usually accomplished through the use of a subroutine with the variable parameter G*. When the subroutine is called, the appropriate G code is assigned to cut the necessary quadrant.

QUESTIONS AND PROBLEMS

1. What is a loop and how is it used?
2. What is the format of a loop?
3. What is a subroutine and how is it used?
4. What are the formats for defining and calling a subroutine?
5. What is nesting?
6. Write a loop to drill a series of five holes equally spaced at 1 in. apart, 1 in. deep starting at (1,1) at a 0.1 in. clearance plane.
7. Write a subroutine to cut the same series of holes given in Problem #6.
8. What is mirroring and how is it used?

FIGURE 10.7
PROBLEM #10

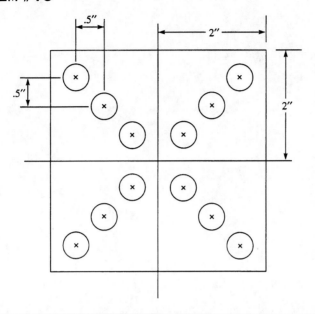

9. Write a short program using a subroutine and mirroring to cut a frame with 3 × 3 × 0.5 in. inside dimensions centered in each quadrant on a 9 × 9-in. workpiece with (0,0) in the center of the workpiece. Use a 0.25-in. diameter end mill.
10. Write a program using a loop, subroutine, and mirroring to drill three equally spaced holes in each of the four quadrants. Remember to set part origin in the center of the workpiece. Fig. 10.7 illustrates the part features.

Computer Control in CNC Programming

Chapter Objectives

After studying this chapter, the student will be able to
- Describe the role of computers in CNC programming.
- Recognize and describe two of the primary computer languages available for CNC programming.
- Describe the function of a postprocessor.
- Describe the motion directions, surfaces, and modifiers in the APT language.
- Describe the major features of the COMPACT II language.
- Describe point-to-point, linear-cut, and circular motion in the APT and COMPACT II languages.
- Recognize the function and major features of CAD/CAM in CNC programming.
- Define the use and function of distributive numerical control and flexible manufacturing systems.
- Describe the function of work cells.

11.1 Introduction

Many simple CNC programs are relatively short. They can be written out in longhand form, typed in manually, and punched rather easily. Therefore, many manufacturers still use manual programming (without the aid of the computer) for these simpler programs.

When the program becomes longer or more complicated, computers can save the programmer many hours of work. The computer is almost 100% accurate, depending on the ability and effort of the programmer. Most errors are the fault of the programmer, not the computer. The computer is only as good and as accurate as the program the programmer enters.

This chapter will describe and give examples of the uses and available languages in CNC programming. Also included in this chapter are applications of the computer in manufacturing such as CAD (computer-aided drafting) and CAM (computer-aided manufacturing). The concepts of flexible manufacturing, DNC, and work cells are provided to enhance your understanding of the practical employment of computer technology in the manufacturing industry.

11.2 ASPECTS OF THE COMPUTER IN CNC

In the last two decades, engineers have capitalized on the power and capabilities of the computer in numerical control applications. This was due, in part, to the advancements in computer technology spurred on by advancements in the electronics industry, such as the reduced cost of PCs and reduced size of components.

Onboard memory allows the part program to be stored in the MCU. The program is input into the MCU once, where it is stored in memory; then the program can be read and executed directly from memory as many times as necessary. Some CNC machines have the ability to store more than one program in memory, depending on available memory and MCU configuration. The ability to store part programs speeds the operation of the CNC machine. Information comes much faster from the computer's memory than it can be read from the tape reader or downloaded from a remote source. Memory storage of programs also reduces the wear and tear on input media, since the program is loaded only once and then stored until needed again. Computers create, transfer, store, and edit part programs, exercising a vital role in modern manufacturing industries.

Computers are complicated tools that perform three basic functions: data input, data processing, and data output. They can perform millions of operations per second, much more rapidly than can be done manually. In addition, they do not require breaks, holidays, sick leave, or vacations. Thus, the computer can be set up to perform automated work unattended. However, a person must program the parts and set up the computer-to-MCU link.

Computers generally create jobs rather than replacing workers. In this sense the computer has been one of the most misunderstood tools created by technology. Computers can be used to save many hours of work and perform hazardous or tedious jobs economically and accurately. However, in order for the computer to perform correctly the programmer and computer must speak the same language. This generally requires additional training or a specialized background.

So far, discussion in this text has centered on the data input and output functions of the computer such as tape preparation or direct downloading of programs. Computers are also able to use languages to translate conversational statements into program code. Computer programming languages extend the power and flexibility of the computer in CNC programming.

11.3 COMPUTER LANGUAGES AVAILABLE FOR CNC PROGRAMMING

Several computer languages with a variety of purposes, focuses and abilities are currently available for CNC programming. Some languages offer a limited vocabulary and few functions, while others offer more features than it would seem a programmer could ever fully utilize. These languages are intended to be conversational and logical in their approach. However, the programmer must be acquainted with the vocabulary the language uses to effectively write a program in that language.

The selection of a CNC language is an involved decision based on the type of parts to be programmed, type of machine(s) used, computer power and type available, part and program features desired, and the costs involved. Most computer language programs for CNC programming fall within the $300 to $30,000 range, depending on the factors listed previously.

These computer languages act as translators. They translate the conversational statements typed in by the programmer into appropriate code. The programmer inputs the original or source program statements through the keyboard of the computer in whatever language available. Several different types of machines may be programmed using the same CNC language, as long as the appropriate translator program is available.

The programmer must tell the computer which machine and tools are used, specifications concerning the part being manufactured and the cutter path that should be taken to manufacture the part. The computer then translates these statements into appropriate code through the use of a separate program called a **postprocessor.**

The postprocessor is a separate computer program that reads the statements input by the programmer, translates these statements into appropriate code, and writes the translated code to an output file. This file may be printed, punched, or downloaded directly to the MCU. Postprocessors are written for specific machines and vary from MCU to MCU. They may also vary depending on the type of computer, machine, and language used.

Two of the more common programming languages for numerical control are **APT** and **COMPACT II**. APT (Automatic Programmed Tool) is the oldest of the CNC languages in general use. It offers the largest vocabulary and substantial power in programming a variety of machines. APT is generally restricted to larger capacity computers, although less powerful versions are available for smaller computers and PCs. APT has the ability to perform complicated calculations for complex curves, tabulated cylinders, and multiaxis programming as well as point-to-point, linear-cut, and contouring applications. APT has generally been accepted as one of the standard CNC languages in the manufacturing industry.

COMPACT II is also widely used, often on an interactive basis through remote terminals. COMPACT II (as well as APT) covers a wide range of applications such as milling, turning, punching, EDM, and flame cutting. COMPACT II offers simple rules and language that make it easy to learn and use. Both languages offer diagnostics during processing of the program, which is helpful in debugging the program before submitting it to the MCU. COMPACT II uses a machine-tool link to postprocess source files. Since APT and COMPACT II are the most common general computer-aided languages used for CNC programming, attention will be focused on them. The following shows a typical APT program and translated code.

```
PARTNO DRILLING EXAMPLE
MACHIN/MILL1,1
PRINT/ON
CLPRNT
SETPT=POINT/-4,0,.1
PT1=POINT/-2,0
PT2=POINT/0,-2
PT3=POINT/2,0
PT4=POINT/0,2
LOADTL/1
CUTTER/.25
SPINDL/1200,RPM
COOLNT/ON
RAPID,FROM/SETPT
CYCLE/DRILL,.5,12,IPM,.1
GOTO/PT1
GOTO/PT2
GOTO/PT3
GOTO/PT4
CYCLE/OFF
RAPID,GOTO/SETPT
END
FINI
```

Translation

```
N5T1M6
N10S1200M41
N15M3
N20M8
N25G81X-2.Y0Z-.5Z.1
N30X0Y-2.F12
```

```
N35X2.Y0
N40X0Y2.
N45G80
N50G0Z.1
N55X-4.Y0
N60M2
```

There are many versions of APT in use today, all differing slightly from the original. These changes are minor and do not significantly affect the basic APT language statements.

The processing of an APT program is divided into four phases. The first phase reads the input file and scans it for errors. Statements are classified and organized according to type of operation. They are then translated into a form ready for the next section. Phase 2 is the arithmetic phase. It receives the information from the first section and, using a library of subroutines, tables, and symbols, generates the equations necessary for describing the cutter path for the part programmed. These equations describe the path generated by the center point of the tool in three-dimensional space. Phase 3 is the edit phase; multiaxis programming, translated cuts, and multiple copy cuts are translated based on output from the second phase. Phase 4 is the postprocessor phase. Based on the MACHIN statement input by the programmer, the proper postprocessor is selected and the data converted into the proper code for the selected MCU. The proper postprocessor must be available for the appropriate code to be generated.

The output from phases 1, 2, and 3 is converted into a *cutter location* or CL file. This describes the center point of the tool in three-dimensional space along the cutter path. A separate program, the postprocessor, is required to convert the CL file into CNC code. The actual program tape or file is a result of postprocessing.

Postprocessors are used to

1. Convert CL file data into machine coordinates.
2. Troubleshoot the program to see that axis travel limits, spindle speeds, feed rates, etc., are not exceeded.
3. Control the tolerance of the part through control of programmed machine movements.
4. Output appropriate G and M codes to control machine movements and functions.
5. Calculate cutter compensation.
6. Output appropriate circular and parabolic interpolated information.
7. Control multiple-axis movements and functions.
8. Produce error diagnostics, as required.

Attempts have been made to standardize postprocessing. One effort is the *binary cutter location,* or BCL, data file. Due to the lack of compatibility between machines, separate postprocessors are required for each different type of MCU. Ideally, the BCL eliminates the need for unique postprocessors, because it is part-oriented rather than machine-oriented.

In this system, postprocessing becomes a function of the machine controller through the BCL program rather than through a separate computer. The BCL program, for BCL-compatible machines, outputs part-oriented information based on the specific MCU's limits and abilities. The original program can be taken to another BCL-equipped machine and postprocessed for its controller. This increases the portability of the program and reduces the need for numerous postprocessors and reprocessing. It also increases the production rate and efficiency of the CNC machines involved.

Currently, there are many postprocessors termed universal. This means that they can be configured and reconfigured to suit the programmer's requirements and specific MCU parameters. Universal postprocessors allow the programmer to specify the appropriate G and M codes output by the postprocessor to coincide with specific MCUs' requirements. In addition, they allow programmers to tailor the output code to suit their needs based on their unique coding input. Universal or configurable postprocessors have been developed for APT and COMPACT II, as well as for many other CNC languages. Programmers must be well acquainted with the hardware and software of the MCU they are programming to properly configure a universal postprocessor.

APT part programming is based on the concept that two planes or surfaces guide the cutter toward a third plane or surface. The two surfaces that guide the cutter along the cutter path are the *drive* and *part surfaces*. The intended destination of the cutter is guided by the *check surface*. Fig. 11.1 shows the relationship between these three surfaces.

Based on program input, the cutter continues along the drive and part surfaces until it reaches the check surface. The cutter may be positioned on, to the left, or to the right of the drive surface while cutting. The cutter may be told to stop TO, ON, or PAST the check surface. Fig. 11.2 illustrates these positions.

Once the cutter has stopped at the check surface, it may be directed to GOLFT, GORGT, GOFWD, GOUP, or GODOWN, corresponding to the intended cutter path. This motion is programmed in relation to the last motion of the cutter. Fig. 11.3 illustrates the result of these motion statements.

In addition, the cutter may be instructed to travel to a specific position by the GOTO statement. Fig. 11.4 illustrates this point-to-point positioning.

Incremental or relative positioning is accomplished through the GODLTA statement. Following the GODLTA statement, the desired

FIGURE 11.1
APT SURFACES

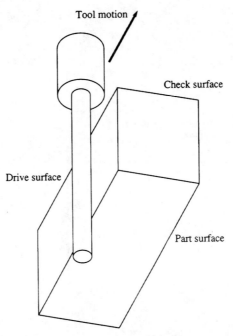

movement in the three axes would be given. Fig. 11.5 on p. 224 is an example of incremental positioning in APT.

Before programming cutter motion, the programmer must define the geometry and positions of the part features by geometry statements that define the entities required for further programming. These are the various points, lines, circles, planes, etc., required for cutter positioning. Fig. 11.6 on p. 225 provides some methods used to define these geometric entities. These entities are given symbols, such as Pn, Ln, and Cn for points, lines, and circles, respectively. A symbol can be any alphanumeric combination of six figures or fewer, with at least one figure being a letter. Fig. 11.7 on p. 226 provides a sample APT source program.

COMPACT II uses reserved symbols for defining entities. DPTn, DLNn, and DCIRn are used to define points, lines, and circles, respectively. In addition, the words MOVE and CUT are used to program rapid moves and movements at the programmed feed rate. Cutter motion is controlled through statements such as CUT,TOLNn, CUT,ONLNn, and CUT,PASTLNn. Fig. 11.8 on p. 227 provides the COMPACT II source program for the same part programmed in Fig. 11.7 for APT.

FIGURE 11.2 ☐☐☐☐☐☐☐☐☐☐☐☐☐☐☐☐☐☐☐☐☐☐☐☐☐
CUTTER POSITION

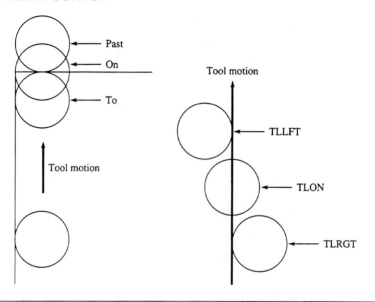

FIGURE 11.3 ☐☐☐☐☐☐☐☐☐☐☐☐☐☐☐☐☐☐☐☐☐☐☐☐☐
CUTTER MOTION

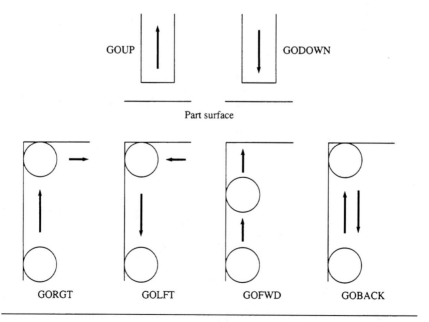

**FIGURE 11.4
GOTO POSITIONING**

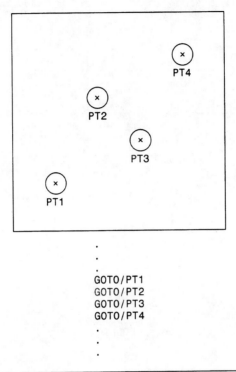

```
    .
    .
    .
GOTO/PT1
GOTO/PT2
GOTO/PT3
GOTO/PT4
    .
    .
    .
```

It is important at this point to note that since the beginning of the CNC industry until the early 1970s, all MCUs were hardware-dependent. This means that the MCU included all of the physical components, such as tape formats, positioning modes, data interpretation, and memory, required to control programming functions. During the 1970s, with the development of more powerful and less expensive electronics, functions that were previously based on hardwired components began to be controlled by computer elements and microprocessors in the MCU.

The switch from hardware dependence to software control appeared at the 1976 Machine Tool Manufacturers' Show. The functions of the MCU were controlled through computer logic, which provided extended capabilities at the same cost. Functions such as recognizing tape formats could be reprogrammed at any time and were not dictated by hardwired components at the time of manufacture. Today, functions can be reprogrammed using punched tape, magnetic tape, floppy disk, or a programmable controller unit. This added flexibility is a major advantage of softwired controllers over their hardwired predecessors. The trend away

FIGURE 11.5 ☐☐☐☐☐☐☐☐☐☐☐☐☐☐☐☐☐☐☐☐☐☐☐☐☐☐☐☐
INCREMENTAL POSITIONING

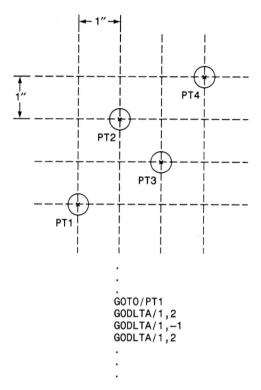

from hardware dependence toward software programming has enhanced the flexibility and power of CNC programming.

11.4 CAD/CAM

The use of the computer in manufacturing has grown considerably over the last two decades. Computer graphics programming such as computer-aided drafting (CAD) and design (CADD) is used extensively. CAD allows the drawing and detailing of parts through the use of the computer. Geometric entities and shapes are drawn and formed. Once the drawing is complete, it may be stored in a central database from which it may be retrieved and processed for manufacturing. CAD also allows for the easy editing of single parts or part families.

FIGURE 11.6
APT GEOMETRY STATEMENTS

```
BASIC POINT DEFINITIONS

P1=POINT/X,Y
P2=POINT/X,Y,Z
P3=POINT/P1,DELTAX,DX,DELTAY,DY,DELTAZ,DZ
P4=POINT/P1,XCOORD,XC,YCOORD,YC,ZCOORD,ZC
P5A=POINT/P1,XYROT,ANGLE
P5B=POINT/P1,YZROT,ANGLE
P5C=POINT/P1,ZXROT,ANGLE
P6A=POINT/RTHETA,XYPLAN,RADIUS,ANGLE
P6B=POINT/THETAR,XYPLAN,ANGLE,RADIUS

BASIC LINE DEFINITIONS

L1=LINE/X1,Y1,X2,Y2
L2=LINE/X1,Y1,Z1,X2,Y2,Z2
L3=LINE/P1,P2
L4A=LINE/XAXIS
L4B=LINE/YAXIS
L5A=LINE/XCOORD,XC
L5B=LINE/YCOORD,YC
L5C=LINE/YCOORD,ZC
L10=LINE/P1,PARLEL,L1
L11=LINE/P1,PERPTO,L1
L12=LINE/PARLEL,L1,XLARGE,XC
L13=LINE/PARLEL,L1,XSMALL,XC
L14=LINE/PARLEL,L1,YLARGE,YC
L15=LINE/PARLEL,L1,YSMALL,YC
L16=LINE/P1,ATANGL,ANGLE
L17=LINE/P1,ATANGL,ANGLE,L1

BASIC CIRCLE DEFINITIONS

C1=CIRCLE/P1,P2,P3
C3=CIRCLE/CENTER,P1,RADIUS,R
C4=CIRCLE/CENTER,P1,TANTO,L1
C5=CIRCLE/CENTER,P1,P2
C6=CIRCLE/CENTER,P1,LARGE,TANTO,C1
C7=CIRCLE/CENTER,P1,SMALL,TANTO,C1
C9=CIRCLE/C1,LARGE,OFFSET
C10=CIRCLE/C1,SMALL,OFFSET

BASIC PLANE DEFINITIONS

PL1=PLANE/1,2,3,4
PL2=PLANE/P1,P2,P3
PL3=PLANE/P1,PARLEL,PL1
PL4=PLANE/P1,PERPTO,PL1,PL2
PL5=PLANE/PERPTO,PL1,P1,P2
PL6A=PLANE/PARLEL,PL1,XLARGE,DX
PL6B=PLANE/PARLEL,PL1,XSMALL,DX
PL6C=PLANE/PARLEL,PL1,YLARGE,DY
PL6D=PLANE/PARLEL,PL1,YSMALL,DY
PL6E=PLANE/PARLEL,PL1,ZLARGE,DZ
PL6F=PLANE/PARLEL,PL1,ZSMALL,DZ
PL7=PLANE/L1
```

FIGURE 11.7
SAMPLE APT PROGRAM

```
%
N5T1M6
N10S3000M21
N15M3
N20G0X-1.5Y0.Z0.
N25G3X-1.Y-.5I.5J0.F25.
N30G1X1.
N35G3X1.5Y0.I0.J.5
N40X1.Y.5I.5J0.
N45G1X-1.
N50G3X-1.5Y0.I0.J.5
N55X-1.4841Y-.125I.5J0.
N60G0X-5.Y0.Z.1
N65M2

MACHIN/MMPOST,33
PRINT/ON
CLPRNT
SETPT=POINT/-5,0,.1
LN1=LINE/XAXIS
LN2=LINE/YAXIS
LN3=LINE/PARLEL,LN2,XSMALL,1
LN4=LINE/PARLEL,LN2,XLARGE,1
LN5=LINE/PARLEL,LN1,YLARGE,.5
LN6=LINE/PARLEL,LN1,YSMALL,.5
PT1=POINT/INTOF,LN3,LN1
PT2=POINT/INTOF,LN4,LN1
CIR1=CIRCLE/CENTER,PT1,RADIUS,.5
CIR2=CIRCLE/CENTER,PT2,RADIUS,.5
PL1=PLANE/0,0,-.25,0
CUTTER/.25
LOADTL/1
SPINDL/3000,RPM
FEDRAT/25,IPM
RAPID,FROM/SETPT
GO/ON,CIR1,ON,PL1,ON,LN1
TLON,GORGT/CIR1,TANTO,LN6
GOFWD/LN6,TANTO,CIR2
GOFWD/CIR2,TANTO,LN5
GOFWD/LN5,TANTO,CIR1
GOFWD/CIR1,PAST,LN1
RAPID,GOTO/SETPT
END
FINI
```

CAD has its basis in the computer. It uses the same basic construction techniques as conventional drafting with the added flexibility of the computer. CAD programs allow for the easy creation, display, editing, and transformation of part drawings. Additionally, stored programs are less cumbersome to deal with, easier to retrieve, and less likely to be misplaced or misfiled. Most CAD drawings should also be available in hardcopy form, just in case the power fails, the computer eats the disk, the computer drive fails, or another unforeseen act occurs. The computer is just a tool, less prone to error than humans, but not infallible.

FIGURE 11.8 ▫▫▫▫▫▫▫▫▫▫▫▫▫▫▫▫▫▫▫▫▫▫▫▫▫▫▫▫
COMPACT II PROGRAM

```
MACHIN, MILL1
IDENT, OBLONG GROOVE
SETUP, (specific machine positioning and limits information)
BASE, XA, YA, ZA
DLN1, XB
DLN2, YB
DLN3, LN2, -1X
DLN4, LN2, 1X
DLN5, LN1, .5Y
DLN6, LN1, -.5Y
DPT1, LN3, LN1
DPT2, LN4, LN1
DCIR1, PT1, .375R
DCIR2, PT2, .375R
DPLN1, -.25ZB
MTCHG, TOOL1, 12IPM, 2000RPM, TD.25
MOVE, ONCIR1, ONLN6
CUT, -.25ZB
CUT, LN4
OCON, CIR1, CCW, S(LOC), F90
CUT, LN3
OCON, CIR2, CCW, S(LOC), F270
MOVE, .5ZB
HOME
STOP
```

In computer-graphics programming, the programmer draws a part using any of several CAD programs currently available. After drawing the part on the computer screen, the programmer traces or programs the cutter path using an appropriate input device such as a mouse, digitizing tablet, or light pen. MCU used, tools, speeds, and feed rates may be defined as needed or in advance. After the geometry and cutter path have been defined, the computer translates this information into appropriate code.

The advantage of the CAD/CAM link is that detailed part geometry does not have to be created through part-definition statements. The geometry exists and cutter paths are easily created, edited, and stored directly. Recreating existing part geometry is not necessary or efficient. Part designs may be recalled and altered as necessary directly. Part programs (cutter paths) may also be recalled and edited, redefined, and reprocessed directly on the part drawing, reducing the chance for error in recreating part geometry. Verification is as easy as following the tool visually on the graphics terminal. Dry runs can be performed on the display terminal rather than tying up valuable production time just to discover that an error in recreating the geometry occurred that could have easily been corrected in a CAM program. CAM programs do not replace the dry-run procedure, but they do reduce the time required for cutter path verification.

This CAD/CAM link provides added power and flexibility to the programmer. Design changes are easily made. Families of parts may be grouped together on the computer. Design tolerances and specifications may be altered on the computer, rather than redrawing and reprogramming the entire part. CAD/CAM together speed the concept-to-production process. Computer-assisted part programming is often much faster than comparable manual work in which geometry and tool paths must be recreated.

11.5 DISTRIBUTIVE NUMERICAL CONTROL AND FLEXIBLE MANUFACTURING

CNC evolved from the direct numerical control (NC) systems of the 1960s and 1970s. These systems did little more than replace the tape reader. The computers required to control direct numerical control systems were expensive, and the entire system collapsed if the computer went down. Tape readers were available as backups in case of failure. These systems were often difficult to coordinate, as each of the NC machines was partially or completely controlled from the central computer. With the expense of the computer and software and problems in coordinating the system, direct numerical control has been considered economically unfeasible for all but the largest manufacturing implementation. A better approach to central control of operations was adopted—distributive numerical control (DNC).

Distributive numerical control evolved during the late 1970s and early 1980s, when central computer systems were used to control several machine tools through intermediary links. Customarily, tape readers and editors were available at the machine site if needed. DNC systems employ a series of networked computers to coordinate a number of CNC machines. This system of computer-to-computer linkage solved some of the earlier problems associated with direct numerical control systems. DNC is also able to collect data on production and machine tool status while providing centralized control of the machine. DNC systems are a major part of what are known as flexible manufacturing systems. Fig. 11.9 shows a schematic of a typical DNC system.

Flexible manufacturing systems (FMS) are gaining greater acceptance among manufacturers. A flexible manufacturing system is based on hardware and software components that act in a coordinated fashion to produce a wide variety of similar parts. A flexible manufacturing system consists of CNC machines, robots, inspection equipment, and material handling equipment that can take a part from raw material, perform all necessary machining, material handling, and inspection to produce a finished part. An FMS is an automated production and assembly line capable of automatic operation without supervision. It is software-based, typically controlled through a DNC system. A typical FMS is shown in Fig. 11.10.

Larger flexible manufacturing systems are often made up of smaller units called *work cells*. A work cell usually contains one or more CNC machines and a robot for part handling. The cell performs the necessary function for which it was designed and the robot or part transfer vehicle moves parts through the cell. The primary element in the work cell and in FMS is the CNC machining or turning center. The automatic tool changing capability of these machines allows the system to run unattended.

FIGURE 11.9 ☐☐☐☐☐☐☐☐☐☐☐☐☐☐☐☐☐☐☐☐☐☐☐☐☐☐
DISTRIBUTIVE NC SYSTEM

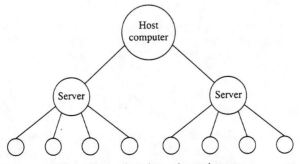

FIGURE 11.10
FMS

(Courtesy of Cincinnati Milacron)

The system is also equipped with monitoring systems that provide feedback on part, tool, and work-flow conditions. Tool monitoring systems in the CNC machine can detect dull, worn, or broken tooling. Sensors can detect whether a part is present and count the number of pieces produced. In addition, an inspection station may be set up to determine whether the part meets specifications. Work cells are widely used in manufacturing today and the use of the flexible manufacturing system is growing. Fig. 11.11 shows a typical work cell.

The FMS offers increased production rates while reducing the amount of work in progress. An FMS frequently offers improved quality and part throughput while reducing inventory and worker requirements. More accurate production scheduling and reduced lead time on orders are accomplished through round-the-clock production, all machine-controlled and software-based. A fully automated FMS requires only cursory supervision and some provision for handling contingencies. Flexible manufacturing systems are becoming necessary in manufacturing industries that intend to keep up with global competition.

11.6 SUMMARY

Computers perform three basic functions: data input, data processing, and data output. Their principal advantages are speed and accuracy, although their accuracy depends on the effort and ability of the programmer.

FIGURE 11.11
WORK CELL

(Courtesy of Cincinnati Milacron)

Several computer languages are available for programming CNC machines. These languages translate conversational statements input by the programmer into appropriate code. The languages use a separate program called a postprocessor.

The CAD/CAM link allows programmers to design and submit the part geometry to a postprocessor in a computer, where appropriate code is produced. This significantly reduces the development time for parts and allows greater ease in design changes.

DNC and FMS systems allow greater flexibility in part programming. However, they only begin to tap into the vast capabilities of the computer in CNC. Distributive NC offers more control over CNC machines while being able to record other manufacturing data concerning production. FMS offers a stand-alone system that, if fully automated, lowers costs by increasing quality and throughput times while reducing the inventory and workers required. Both of these systems have hardware and software components. Tomorrow's software will extend the capabilities of these systems and further tap the enormous power of computer control.

QUESTIONS AND PROBLEMS

1. What are the advantages of using the computer in CNC programming?
2. What three functions are basic to all computers used in numerical control?
3. List the two primary CNC languages commonly available for part programming.
4. What is a CL file?
5. What is a postprocessor?
6. What does a postprocessor do?
7. Describe the four phases of APT programming.
8. Define CAD/CAM.
9. What advantages do CAD/CAM links offer to CNC programming?
10. What is distributive NC?
11. Describe a flexible manufacturing system.
12. What is a work cell?

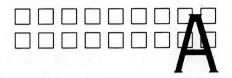

COMMONLY USED NATIONAL CODES

Character	Significance
A	Angular dimension around the X axis
B	Angular dimension around the Y axis
C	Angular dimension around the Z axis
D	Angular dimension around a special axis, third feed function, or tool offset
E	Angular dimension around a special axis or second feed function
F	Feed rate
G	Preparatory function
H	Tool compensation
I	Used to program arc centerpoint in the X axis for circular interpolation
J	Used to program arc centerpoint in the Y axis for circular interpolation
K	Used to program arc centerpoint in the Z axis for circular interpolation
L	Threading
M	Miscellaneous function
N	Sequence number
O	Some systems use O in place of N for sequence number
P	Third rapid traverse code or tertiary motion dimension parallel to the X axis
Q	Second rapid traverse code or tertiary motion dimension parallel to the Y axis
R	First rapid traverse code, tertiary motion dimension parallel to the Z axis, the radius for constant surface speed calculations, or radius designation in circular interpolation
S	Spindle speed
T	Tool number; some systems combine tool number and offset register
U	Secondary motion dimension parallel to the X axis
V	Secondary motion dimension parallel to the Y axis
W	Secondary motion dimension parallel to the Z axis
X	Primary motion dimension in the X axis

Y	Primary motion dimension in the Y axis
Z	Primary motion dimension in the Z axis
%	Rewind stop code
/	Block delete

G Codes

Code	Description
G00	Rapid traverse positioning
G01	Linear interpolation
G02	Circular interpolation in the CW direction
G03	Circular interpolation in the CCW direction
G04	Dwell
G05	Unassigned
G06	Parabolic interpolation
G07	Unassigned
G08	Acceleration override
G09	Deceleration override
G10	Tool length offset value
G11–12	Unassigned
G13–16	Axis selection
G17	Selects X-Y plane
G18	Selects Z-X plane
G19	Selects Y-Z plane
G20	Inch data input (some systems)
G21	Metric data input (some systems)
G22	Safety zone specification
G23	Safety zone override
G24–26	Unassigned
G27	Reference point return check
G28	Return to reference point
G29	Return from reference point
G30	Return to second reference point
G31–32	Unassigned
G33	Threading, constant lead
G34	Threading, increasing lead
G35	Threading, decreasing lead
G36–39	Unassigned
G40	Cutter diameter compensation cancel or tool nose radius compensation cancel (lathe programming)
G41	Cutter diameter compensation left or tool nose radius compensation left (lathe programming)
G42	Cutter diameter compensation right or tool nose radius compensation right (lathe programming)
G43	Tool length compensation (+) direction
G44	Tool length compensation (−) direction
G45	Toll offset increase
G46	Tool offset decrease
G47	Tool offset double increase
G48	Tool offset double decrease
G49	Tool length compensation cancel

Code	Description
G50	Scaling off or loop end
G51	Scaling on or loop start
G52–69	Unassigned
G70	Inch data input (some systems) or finish turning cycle (lathe programming)
G71	Metric data input (some systems) or rough turning cycle (lathe programming)
G72	Facing cycle (lathe programming)
G73	Peck-drilling cycle or pattern repeat (lathe programming)
G74	Countertapping cycle or Z-axis peck-drilling (lathe programming)
G75	Unassigned or groove cutting (lathe programming)
G76	Fine boring cycle or multipass threading (lathe)
G77–79	Unassigned
G80	Canned cycle cancel
G81	Drilling cycle
G82	Counterboring cycle
G83	Peck drilling cycle
G84	Tapping cycle
G85	Boring cycle with feed return to reference level
G86	Boring cycle with rapid return to reference level
G87	Back boring cycle
G88	Boring cycle with manual return
G89	Boring cycle with dwell before feed return
G90	Absolute positioning
G91	Incremental positioning
G92	Preset zero reference point
G93	Inverse time feed rate
G94	IPM feed rate
G95	IPR feed rate
G96	Constant surface speed in feet per minute
G97	Constant spindle speed in RPM
G98	Return to initial level
G99	Return to reference level

Miscellaneous Functions

Code	Description
M00	Programmed stop
M01	Optional stop
M02	End of program with tape rewind
M03	Spindle start CW
M04	Spindle start CCW
M05	Spindle stop
M06	Tool change
M07	Coolant #2 on
M08	Coolant #1 on
M09	Coolant off
M10	Clamp
M11	Unclamp
M12	Unassigned

M13	Spindle on CW, coolant on
M14	Spindle on CCW, coolant on
M15	Rapid traverse or feed in the (+) direction
M16	Rapid traverse or feed in the (−) direction
M17	Spindle and coolant off
M18	Unassigned
M19	Oriented spindle stop
M20	Unassigned
M21	Mirror X axis
M22	Mirror Y axis
M23	Mirror imaging off
M24–29	Unassigned
M30	End of program, memory reset
M31	Interlock bypass
M32–39	Unassigned
M40–45	Select speed range
M46–47	Unassigned
M48	Override cancel off
M49	Override cancel on
M50–97	Unassigned
M98	Jump to subroutine
M99	Return from subroutine

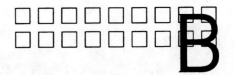

Useful Formulas and Tables

Circumference of a circle = $\pi \cdot$ diameter (d)

Area of a circle = $\pi \cdot r^2$

Surface area of a sphere = $\pi \cdot d^2$

Volume of a sphere = $\dfrac{\pi \cdot d^3}{6}$

$c^2 = a^2 + b^2$

Use Fig. B.1 with the following formulas.

$$\text{Sine} = \frac{\text{Opposite}}{\text{Hypotenuse}} \qquad \text{Cosecant} = \frac{\text{Hypotenuse}}{\text{Opposite}}$$

$$\text{Cosine} = \frac{\text{Adjacent}}{\text{Hypotenuse}} \qquad \text{Secant} = \frac{\text{Hypotenuse}}{\text{Adjacent}}$$

$$\text{Tangent} = \frac{\text{Opposite}}{\text{Adjacent}} \qquad \text{Cotangent} = \frac{\text{Adjacent}}{\text{Opposite}}$$

FIGURE B.1
ANGLES AND SIDES OF A RIGHT TRIANGLE

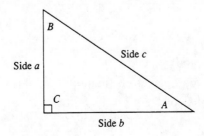

$$\text{Cutting speed (FPM)} = \frac{\text{RPM} \cdot \text{Diameter (in)} \cdot \pi}{12 \text{ in/ft}}$$

where RPM is the spindle speed in revolutions per minute and the diameter is the diameter of the cutter in inches.

An approximation of this formula is

$$\text{Cutting speed (FPM)} = \frac{\text{RPM} \cdot \text{Diameter (in)}}{4}$$

where 4 is an approximation of $12/\pi$.

In the metric system

$$\text{Cutting speed (m/min)} = \frac{\text{RPM} \cdot \text{Diameter (mm)} \cdot \pi}{1000 \text{ mm/m}}$$

An approximation of this formula is

$$\text{Cutting speed (m/min)} = \frac{\text{RPM} \cdot \text{Diameter (mm)}}{300}$$

where 300 is an approximation of $1000/\pi$.

$$\text{Spindle RPM} = \frac{\text{Cutting speed} \cdot 12 \text{ in/ft}}{\text{Diameter (in)} \cdot \pi}$$

where the cutting speed in surface feet per minute is found in the tables and the diameter is taken from the part or cutter.

An approximation of this formula is

$$\text{Spindle RPM} = \frac{\text{Cutting speed} \cdot 4}{\text{Diameter (in)}}$$

$$\text{Milling feed rate (IPM)} = \text{RPM} \cdot T \cdot N$$

or

$$\text{Feed (mm/min)} = (\text{mm/tooth chip load}) \cdot \text{RPM} \cdot N$$

where RPM is the spindle speed.
T is the chip load per tooth found in the tables.
N is the number of teeth on the cutter.

$$\text{Lathe feed rate (IPR)} = \frac{\text{IPM}}{\text{RPM}}$$

or

$$\text{Feed (IPR)} = \frac{\text{(mm/min)}}{\text{RPM}}$$

where IPM is the feed rate in inches per minute.
RPM is the spindle speed.

$$\text{Lathe feed rate (IPM)} = \text{RPM} \cdot \text{IPR}$$

or

$$\text{Feed (mm/min)} = \text{RPM} \cdot \text{(mm/rev)}$$

where IPR is the feed rate in inches per revolution. This formula is used for drills, reamers, countersinks, and lathe programming.

$$\text{Lead of a thread} = P \cdot I$$

where P is the thread pitch.
I is the number of leads.

$$\text{Thread pitch} = \frac{1}{N}$$

where N is the number of threads per inch (TPI).

For threading or tapping the feed rate =

$$\text{Lead in inches} \cdot \text{RPM} = \text{RPM/TPI}$$

or

$$\text{Feed (mm/min)} = \text{RPM} \cdot \text{Pitch}$$

CUTTING SPEEDS (FEET PER MINUTE)

Material	Drilling	Milling	Reaming	Tapping	Boring
Aluminum	200	300	175	90	300
Brass-soft	145	200	125	100	150
Brass-hard	125	175	100	75	125
Bronze	135	90	50	45	75
Cast iron	60	50	45	35	90

Cutting Speeds (feet per minute) (Continued)

Material	Drilling	Milling	Reaming	Tapping	Boring
Copper	75	100	50	40	100
Magnesium	250	400	180	150	400
Monel	50	75	35	25	50
Mild steel	95	125	50	45	75
Stainless steel	55	45	35	25	50
Titanium	50	50	45	30	60
Tool steel	40	75	30	20	45

Note: These values are for High Speed Steel (HSS) cutters. For carbide cutters, these values may typically be doubled. Milling values are for finishing cuts. Roughing cuts are approximately 50–75% of finish values.

Feed Rate

Drill Size (in)	Feed Rate (IPR)
<0.125	0.001–0.002
0.125–0.250	0.002–0.004
0.250–0.500	0.004–0.007
0.500–1.000	0.007–0.015
>1.000	0.25

Chip Load per Tooth for Milling

Material	Face Mill	Side Mill	End Mill
Aluminum	0.020	0.012	0.010
Brass	0.013	0.008	0.006
Bronze	0.012	0.008	0.006
Cast iron	0.012	0.006	0.006
Low carbon steel	0.010	0.005	0.005
Medium carbon steel	0.009	0.005	0.004
High carbon steel	0.006	0.003	0.002
Stainless steel	0.006	0.004	0.002

Note: These are conservative, recommended values and are not absolute. Specific machining conditions may dictate higher or lower values.

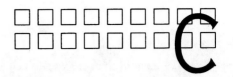

General Safety Procedures Relating to CNC Operations

These procedures are a guide in working with and around manufacturing equipment. By necessity, they are general in nature. The best safety procedure is to use good common sense. If you have a question, ask. Do not assume that things will work out for the best; they usually do not and someone gets hurt.

1. Wear safety glasses at all times when around machinery.
2. Wear proper clothing. Avoid loose articles, such as jewelry, neckties, and gloves, that dangle or could be caught in the machines.
3. Wear safety shoes.
4. Keep longer hair covered and tied back when working with machinery.
5. Practice good housekeeping. Keep the work area clean and clear of obstructions.
6. Clean up any oil, grease, and spills immediately.
7. Do not use compressed air to blow chips off the machine. This may injure someone close to the machine.
8. Avoid grinding or material-removal operations that may result in dust, dirt, or chips on the machine, causing undue wear.
9. No horseplay. Take your work seriously.
10. Do not lay tools, parts, or materials on the machine. It is not a workbench.
11. Use proper lifting techniques when loading and unloading the machine. Use the legs, not the back.
12. Keep hands and face away from the spindle when it is moving.
13. Make all setup, loading, and unloading operations with the spindle fully stopped.

14. Make certain that the workpiece is firmly clamped before executing the program.
15. Use a cloth or a pair of gloves when handling cutters or tools. They are sharp.
16. Replace dull or broken tools immediately. Store them in the proper place so that someone else does not try to use them.
17. Do not attempt to run the machine before being properly instructed on each of the controls, including the stop and emergency stop functions.
18. Make sure that all guards and safety devices are in position and working properly.
19. Maintenance work should be performed by qualified personnel. If other than routine maintenance is required, let people who are trained for the work do it.
20. Always dry-run a new program first. This will help in diagnosing or avoiding problems.
21. Become acquainted with the particular machine and controller combination. This includes programming particulars, controls, overrides, functions, and capabilities.
22. Do not try to remove chips by hand. Keep chips cleared while the spindle is stopped using a brush.
23. Respect the machine. It can be dangerous.
24. Use your head. The instructors are there to help you. Rely on their knowledge and experience. CNC operations can be safe and fun if you take the time to do them right.

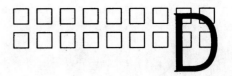

Sample Part Programs

EXAMPLE D.1 ☐☐☐☐☐☐☐☐☐☐☐☐☐☐☐☐☐☐☐☐☐☐☐☐☐☐☐☐

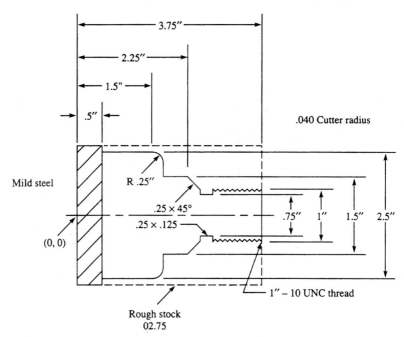

```
%
/G28 U0 W0                          (Return to Machine Zero)
/G0 U-2 W-1                         (Rapid to Intermediate Point)
/G92 X7 Z5                          (Preset Absolute Zero)
N10 G95 G96 F.006 S350 T101 M13     (Constant Surface Speed, IPR Feed
                                    Rate, Call Tool #1 and Offset #1,
                                    Turn Chuck ON CW rotation, Coolant
                                    ON)
N20 G0 G90 X2.85 Z3.85              (Rapid to Clearance in X and Z)
N30 G71 P40 Q140 U.01 W.005 D.020   (Rough Turning Cycle)
N40 G0 X0                           (Rapid to Part Centerline)
N50 G1 Z3.75                        (Feed into Part)
N60 X1                              (Face Out to First Diameter)
N70 Z2.5                            (Turn to Length)
N80 X1.2708                         (Feed to Start of Chamfer)
N90 X1.5 Z2.2708                    (Cut Chamfer)
N100 Z1.75                          (Turn to Length)
N110 X2                             (Second Diameter)
N120 G3 X2.5 Z1.5 I0 K.25           (CCW Circular Arc)
N130 G1 Z.5                         (Finish Length)
N140 X2.85                          (Return to Clearance)
N150 G70 P40 Q140                   (Finish Turning Cycle)
N160 G0 X5 Z4 T202                  (Rapid to Intermediate Point, Call
                                    Grooving Tool and Offset #2)
N170 F6 S200                        (Set Feed and Speed for Grooving)
N180 X1.1 Z2.5                      (Rapid to Clearance)
N190 G1 X.75                        (Feed to Depth)
N200 G0 X1.1                        (Rapid to Clearance)
N210 X5 Z4 T303                     (Return to Intermediate Point, Call
                                    Threading Tool and Offset #3)
N220 X1 Z3.85                       (Position at Clearance)
N230 G95 G97 F.1 S500               (IPR Feed Rate, Constant RPM)
N240 G76 X.87 Z1 I0 K.065 D.020 F.1 (Multiple-Pass Threading Cycle)
N250 G0 X7 Z5 T300 M30              (Return to Intermediate Point, End
                                    Program, Cancel Offset)
%
```

Example D.2

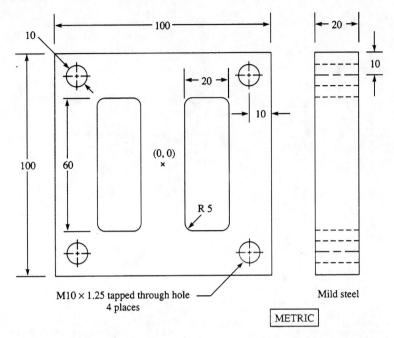

M10 × 1.25 tapped through hole
4 places

Mild steel

METRIC

```
#1
X-15 Y0 Z3                              (Position over Left Slot)
G1 Z*                                   (Feed to depth [variable assigned
                                         when calling macro])
                                        (Cut statements required for slot)
Y25
X-25
Y-25
X-15
Y0
$
%
G0 G71 G90 X-100 Y0 F300 S1200 T1 M6    (Load 8.70⌀ Drill @ TC/PC)
G81 X-40 Y40 Z3 Z25                     (Drill Cycle, Upper Left)
X40 Y40                                 (Upper Right)
X40 Y-40                                (Lower Right)
X-40 Y-40                               (Lower Left)
G80                                     (Cycle Cancel)
G0 G90 X-100 Y0 F250 S200 T2 M6         (Load 10 x 1.25 Tap @ TC/PC)
G84 X-40 Y40 Z5 Z25                     (Tapping Cycle, Upper Left)
X40 Y40                                 (Upper Right)
X40 Y-40                                (Lower Right)
X-40 Y-40                               (Lower Left)
G80                                     (Cycle Cancel)
G0 G90 X-100 Y0 F250 S1000 T3 M6        (Load 10⌀ Flat End Mill @ TC/PC)
=#1 Z*-10                               (Call Macro #1 for Left Slot with
                                         Depth Variable set at -10)
=#1 Z*-20                               (Call Macro #1 for Left Slot with
                                         Depth Variable set at -20)
G0 Z3                                   (Rapid to clearance)
X0 Y0                                   (Position over Part Center)
G31                                     (Mirror X values)
=#1 Z*-10                               (Call Macro #1 for Right Slot with
                                         Depth Variable set at -10)
=#1 Z*-20                               (Call Macro #1 for Right Slot with
                                         Depth Variable set at -20)
G0 Z3                                   (Rapid to clearance)
X-100 Y0 M2                             (Position at TC/PC, End Program)
%
```

EXAMPLE D.3

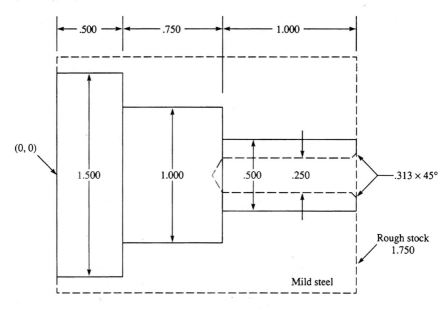

```
%
/G28 U0 W0                              (Return to Machine Zero)
/G0 U-2 W-1                             (Position at Intermediate Point)
/G92 X7 Z5                              (Preset Absolute Zero)
N10 G95 G96 F.006 S350 T101 M13         (Constant Surface Speed, IPR Feed
                                         Rate, Call Tool #1 and Offset #1,
                                         Turn Chuck On CW, Coolant ON)
N20 G0 G90 X1.85 Z2.25                  (Position at Clearance for Facing)
N30 G1 X0                               (Face Part)
N40 G0 Z2.35                            (Rapid to Z Clearance)
N50 X1.85                               (Rapid Return to X Clearance)
N60 G71 P60 Q140 U.05 W.05 D.05         (Rough Turning Cycle)
N70 G0 X0                               (Position for Facing Out)
N80 G1 Z2.25                            (Feed Into Part)
N90 X.5                                 (First Diameter)
N100 Z1.25                              (Turn to Length)
N110 X1                                 (Second Diameter)
N120 Z.5                                (Turn to Length)
N130 X1.5                               (Third Diameter)
N140 Z0                                 (Finish Length)
N150 G70 P60 Q140 F.003                 (Finish Turning Cycle)
N160 G0 G90 X5 Z4 T202                  (3/4 Spot Drill)
N170 G94 G97 F6 S600                    (Constant RPM, IPM Feed Rate)
N180 X0 Z2.35                           (Position at Clearance)
N190 G1 Z2.0937                         (Drill to Depth)
N200 G0 Z2.35                           (Rapid to Clearance)
N210 X5 Z4 T303                         (.25 Drill)
N220 F16 S1600                          (Constant RPM, IPM Feed Rate)
N230 G0 X0 Z2.35                        (Position at Clearance)
N240 G1 Z1.75                           (First Increment)
N250 G0 Z1.85                           (Rapid to Clearance)
N260 G1 Z1.5                            (Second Increment)
N270 G0 Z1.6                            (Rapid to Clearance)
N280 G1 Z1.1778                         (Drill to Depth)
N290 G0 Z2.35                           (Rapid Out of Part)
N300 X7 Z5 T300 M30                     (Return to Intermediate Point,
                                         Cancel Offset, End of Program)
%
```

Example D.4

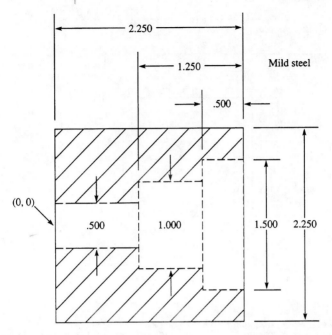

```
%
/G28 U0 W0                              (Return to Machine Zero)
/G0 U-2 W-1                             (Position at Intermediate Point)
/G92 X7 Z5                              (Preset Absolute Zero)
N10 G94 G97 F8.5 S850 T101 M13          (Constant RPM, IPM Feed Rate, Call
                                        Tool #1 and Offset #1, Turn Chuck
                                        On CW, Coolant ON, 15/32 Drill)

N20 G0 G90 X0 Z2.35                     (Rapid to Clearance)
N30 G1 Z1.75                            (First Increment)
N40 G0 Z1.85                            (Rapid to Clearance)
N50 G1 Z1.5                             (Second Increment)
N60 G0 Z1.6                             (Rapid to Clearance)
N70 G1 Z1.1778                          (Drill to Depth)
N80 G0 Z2.35                            (Rapid to Clearance)
N90 X5 Z4 T202                          (3/8 Boring Bar)
N100 G95 G96 F.006 S350                 (Constant Surface Speed, IPR Feed
                                        Rate)

N110 G0 X.45 Z2.35                      (Position at Clearance)
N120 G71 P130 Q180 U-.010 W-.005 D.020  (Rough Turning Cycle)
N130 G0 X1.5                            (First Diameter)
N140 G1 Z1.75                           (Turn to Length)
N150 X1                                 (Second Diameter)
N160 Z1                                 (Turn to Length)
N170 X.5                                (Third Diameter)
N180 Z0                                 (Turn to Finish Length)
N190 G70 P130 Q180 F.003                (Finish Turning Cycle)
N200 X7 Z5 T200 M30                     (Return to Intermediate Point,
                                        Cancel Offset, End of Program)
%
```

Example D.5

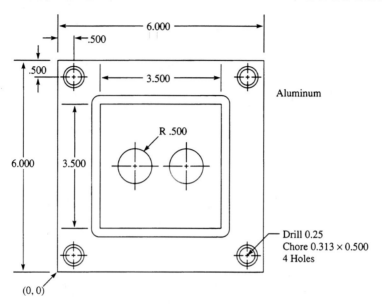

```
%                                      (Rewind Stop Code)
G0 G90 X-2 Y0 F16 S1600 T1 M6          (Load .25 Drill at TC/PC)
G81 X.5 Y.5 Z.1 Z1.25                  (Drill Cycle, Lower Left Hole)
X5.5 Y.5                               (Lower Right)
X5.5 Y5.5                              (Upper Right)
X.5 Y5.5                               (Upper Left)
G80                                    (Cycle Cancel)
G0 G90 X-2 Y0 F12 S1200 T2 M6          (Load .313 Counterbore @ TC/PC)
G82 X.5 Y.5 Z.150                      (0.050 deep Spotfacing Cycle)
X5.5 Y.5                               (Lower Right Hole)
X5.5 Y5.5                              (Upper Right)
X.5 Y5.5                               (Upper Left)
G80                                    (Cycle Cancel)
G0 G90 X-2 Y0 F16 S1600 T3 M6          (Load .25 Flat End Mill @ TC/PC)
X1.25 Y1.25 Z.1                        (Position over Groove at Clearance)
G1 Z-.25                               (Feed to depth)
G91 X3.75                              (Incrementally cut along bottom)
Y3.75                                  (Right Side)
X-3.75                                 (Top)
Y-3.75                                 (Left Side)
G0 G90 X-2 Y0 F9 S800 T4 M6            (Load .5 Flat End Mill @ TC/PC)
X2.25 Y2.25                            (Position over left recess)
G1 Z-.5                                (Feed to depth)
G79 J.25 F9                            (Internal Hole Mill Cycle @ R.25)
G0 Z.1                                 (Rapid to Clearance)
X3.75                                  (Position over right recess)
G1 Z-.5                                (Feed to depth)
G79 J.25 F9                            (Internal Hole Mill Cycle @ R.25)
G0 Z.1                                 (Rapid to Clearance)
X-2 Y0 M2                              (Rapid to TC/PC, End Program)
%                                      (Rewind Stop Code)
```

Glossary

The following terms are related to or associated with CNC programming and technology. Many of these terms have been used throughout this text and should provide an aid to understanding the material. In addition, some terms have been included to aid the student in understanding current literature and discussion about CNC technology.

A axis Machine axis that provides angular rotation around the X axis. Commonly used with machine accessories such as a rotating head.

Absolute programming Programmed dimensions are specified from an origin or programmed zero reference point. The preparatory function G90 is used to specify that absolute programming will be used.

Accuracy The trueness of the measured value in relation to some known value. For example, the difference between the current tool position and the programmed tool position.

Adaptive control Control system that allows monitoring, comparing, and adjusting machine parameters to optimize machine performance conditions.

Address A symbol (usually a letter) used to denote significant program information or a memory location where this information is stored.

APT Automatic programmed tool. A CNC programming language that allows the definition of geometry, tooling, tool motion, and general machine functions through conversational statements. Requires a postprocessor to translate APT source statements into appropriate CNC code.

Arc Circular tool travel made in the clockwise or counterclockwise direction in one of the plane pairs.

ASCII American Standard Code for Information Interchange. One type of programming code used in CNC tapes.

Auxiliary function A programmable machine function that does not involve cutter positioning or control.

Axis A general direction of machine travel. Positive and negative values are assigned based on an origin or programmed zero reference point.

B axis Machine axis that provides angular rotation about the Y axis. Can be assigned to the rotary head assembly on a vertical milling machine or to various machine accessories.

Backlash Lost motion between mechanical parts due to looseness or wear.

BASE Program statement used in many CNC programming languages to establish the origin or programmed zero reference point.

Binary coded decimal (BCD) A coding system that allows binary representation of alphanumeric characters and numbers.

Binary numbering system Base two numbering system that represents numbers as 1s and 0s.

Bit A binary digit that has only two states, 1 or 0, on or off, presence or absence. This includes holes or the absence of holes in punched tapes and magnetized spots on magnetic media.

Block One programmed line of CNC instruction code.

Block delete The slash "/" code represents the block delete function in CNC programming. When the Block Delete function is enabled by the operator, it allows the MCU to ignore a programmed block of code.

Byte Eight adjacent bits. Bits are generally operated on in computer systems as bytes. The computer also operates on words that usually contain sixteen or thirty-two bits, two or four bytes.

C axis Machine axis that provides angular rotation about the Z axis. Can be assigned to machine accessories such as a rotary head assembly. Can also be found on CNC lathes.

CAD Computer-aided drafting.

CADD Computer-aided drafting/design.

CAM Computer-aided manufacturing.

Canned cycle Programmed sequence of tool motions that are built into the MCU. These allow single-line programming of common operations such as drilling, tapping, and boring. Reduces the amount of programmed code needed to accomplish these functions.

Cartesian coordinate system A system used to define distances with respect to one or more axes. CNC programming is based on this system of coordinates.

Circular interpolation A mode of machine operation that allows simultaneous control of two axes to maintain tool travel along an arc. The preparatory functions G02 and G03 are used to initiate clockwise and counterclockwise motion, respectively.

Closed-loop control systems Machine control system that provides positioning feedback. This allows the detection of positioning errors.

CNC Computer numerical control; the control of machines and machine tools through the use of an on-board computer.

Coding system A system used to describe the formation of characters on a programming medium so that the characters may be understood by the sender and receiver of the information.

COMPACT II A widely used programming language for CNC.

Constant surface speed (CSS) or **Constant cutting speed** Situation specified by an appropriate preparatory function where the spindle speed is adjusted relative to the tool position. The speed increases indirectly with the distance the tool is from the spindle centerline.

Continuous path   Operation performed under continuous numerical control of motion rate and direction.

Contouring   Cutter path that results from the coordinated, simultaneous control of two or more axes.

CRT   Cathode ray tube. Used to display alphanumeric characters and graphics on a fluorescent screen by the controlled emission of an electron beam; the familiar computer screen.

Cutter diameter compensation   When programmed, provides an allowance for the cutter diameter or radius as a vector distance from the part surface to be cut. This prevents the cutter from over- or undercutting the part feature and allows programming of the difference in actual versus programmed cutter diameters.

Cutter offset   The distance from the part surface to the axial center of the cutter.

Cutting speed   The rotational speed of the cutting tool or workpiece. Referred to as revolutions per minute (RPM) or surface feet per minute (SFM).

Cycle time   The amount of time required to complete the set of programmed instructions.

Direct numerical control   Eliminates the need for an intermediary between computer and machine. Data are fed directly from large, remote computers.

Distributive numerical control   Communications between a host computer and the MCU through a distributed network. Also provides data sharing and collection of manufacturing-related information.

Dwell   A programmed interruption in program execution.

EIA   Electronic Industries Association, whose standards RS–244 and RS–358 govern coding systems used in punching tapes for CNC programs.

EOB (End of Block) code   Special character that signifies the end of a block of programmed code. Correlates to the carriage return key on modern equipment.

F address   Used to specify feed rates in CNC programming.

Fixed block format   Early tape coding format that required data be programmed in a particular sequence.

Floating zero   The ability to reprogram the zero reference point.

G codes   Preparatory functions that provide programming of machine tool functions such as motion at programmed feed rate (G01), motion at rapid traverse rate (G00), dwells (G04), absolute programming (G91), and incremental programming (G91).

I address   Secondary motion dimension word for the X axis. Used to program the arc centerpoint in circular interpolation.

Incremental programming   Also called relative positioning. Programmed dimensions are based on the previous tool motion. The incremental distance from the current tool position to the next tool position is programmed relative to the last programmed movement.

IPM   Inches per minute. Used to designate that programmed feed rates are based on time rather than spindle revolutions (IPR).

IPR   Inches per revolution. Used to designate that programmed feed rate values are based on spindle revolutions.

J address Secondary motion dimension word for the Y axis. Used to program the arc centerpoint in circular interpolation.

Jog function Control system feature that allows the operator to manually position the machine table without executing programmed code.

K address Secondary motion dimension word for the Z axis. Used to program the arc centerpoint in circular interpolation. Also used on the lathe or turning center to specify such items as thread pitch in canned threading cycles.

Leading zero suppression MCU feature that eliminates the necessity of programming zeroes to the left of the first significant digit in a coordinate value.

Linear interpolation Mode of machine tool operation that provides straight-line cuts at any angle by simultaneous control of two or more axes. The G01 code initiates linear interpolation at the programmed feed rate.

Looping Ability to repeat specified programmed blocks a number of times.

M functions Used to program miscellaneous functions generally not associated with machine tool travel such as coolant, tool changes, and spindle directions.

Machining centers CNC machine tools capable of performing multiple operations on multiple part faces. Usually equipped with automatic tool changers.

Macro *See* Subroutine.

Manual part programming The coding of programs without the use of the computer.

Manuscript A written or printed copy of the input program in symbolic form.

MCU Machine control unit or machine controller. Contains all the necessary memory, computational equipment, and machine control hardware for the machine to execute part programs. May also contain some limited software such as diagnostics, graphics, and data collection features.

MDI Manual data input feature, which allows the manual entry of programmed information through a keypad on the machine.

Modal commands Commands such as the G90 code for absolute programming that remain in effect after they are programmed until canceled by another command. This allows the programmer to specify a modal value and not have to repeat the command in every block.

N address Block sequence numbers. Normally used to aid the programmer in reading the program. Most MCUs do not require block sequence numbers.

NC Numerical control of machine tools without the use of a computer. Also used as a generic term to include all machines and processes controlled by numerically coded instructions.

Offset Method of compensating for differences in tool lengths and diameters. Tool length offsets are input by the operator prior to program execution. The tool length offset is the distance from the tip of the tool to the surface of the workpiece in the Z axis with the tool fully retracted.

Open-loop control system System of machine control that does not provide positioning feedback and therefore cannot detect positioning errors.

Optional stop OPSTOP. The M01 function is used to program optional stops. Optional stops allow the operator to interrupt program execution to turn the part around, change fixture features, and so on, and resume program execution. The feature works if the operator turns on the OPSTOP function on the MCU key panel.

Origin The zero reference point on a Cartesian coordinate system.

Override control Provides operator override of programmed values such as feed rate and spindle speeds. This option must be enabled before overriding the programmed values.

Parity A method of checking for errors between sender and receiver. In CNC programming a separate bit is used to maintain parity. Depending on the character sent, an additional bit of information may be added to maintain either an even or odd number of bits in the character.

Part program A set of coded instructions written in one of a number of source computer languages or in binary form for manual programming for the purpose of producing parts on NC or CNC machines.

Point-to-point control Machine control that designates specific points for tool positioning with no cutter path control between designated points. Used in limited situations such as drilling.

Polar coordinate system Magnitudes and directions are given for points in relation to a fixed point or pole. These values are vectors.

Postprocessor Separate computer program that translates or converts a cutter location data file into a machine-usable form. The cutter location file is created by the main program processor.

R address Used to specify the clearance plane on certain canned cycles or to program the arc radius in circular interpolation using the radius method.

Rapid traverse Programmed movement at the maximum rate of machine travel. Usually programmed using the G00 preparatory function.

Rectangular coordinates Coordinates based on a fixed origin such as a point in the Cartesian coordinate system.

Register Internal memory locations that provide for the temporary storage of data.

Reliability The ability of a machine to produce accurate parts every time.

Repeatability The closeness or agreement between successive measurements of the same characteristic, using the same methods under identical conditions.

Resolution The smallest increment along an axis that the MCU can control. Common resolutions are 0.001, 0.0001, or 0.00001 in. [0.02, 0.002, 0.0002 mm].

Rotary motion Angular or radial motion along or about a machine axis.

RPM Revolution per minute.

S address Used to program spindle speed.

Sequence number Block identification number associated with a programmed statement.

Significant digit Any number that must be kept or retained to preserve the accuracy or precision of the quantity.

Spindle speed The number of revolutions per minute that the cutting spindle of the machine makes.

Subroutine A separate program within the main program. It allows repetitive operations to be programmed once and repeated a number of times. Subroutines also allow the programming of variable parameters. The parameters and subroutine number are given before executing the macro subroutine.

Surface feet per minute (SFM or SFPM) Term used to designate the feed rate of a machine in relation to the spindle speed. The MCU adjusts the spindle speed to maintain the specified cutting rate. SFM is used primarily on turning machines.

T address Used to specify tool numbers and tool offset registers. Commonly used with the miscellaneous function M06, which specifies that a tool change is to occur.

Tab sequential format Early punched-tape format that used tab codes to specify the word addresses of coded data. The position of the tab code determined which word address was implied. This format was prone to error but reduced the amount of typing required compared to the fixed sequential format.

Tool centerline programming Allows the programmer to program the tool radius centerpoint. Requires the programming of tool radius compensation. Commonly used on turning machines.

Tool length compensation When programmed with the appropriate code words, allows the MCU to access the tool length offset registers where compensation values have been entered.

Tool tip- or edge-programming Allows the programmer to program the tool edge without radius consideration. Commonly used on turning machines for such operations as threading.

Toughness The work per unit volume needed to fracture a material.

Trailing zero suppression Allows the programmer to omit zeroes to the right of the last significant digit in coordinate values.

U address Code word used in some canned cycles or incremental movements along the X axis on some turning machines.

Variable word address or **interchangeable format** Allows the programmer to interchange the word order within a block of programmed code. In addition, many systems do not require unchanged values or zero values to be reprogrammed.

W address Code word used in some canned cycles or incremental movements along the Z axis on some turning machines.

Wear resistance A material's opposition to fracture in application. *See* Toughness.

Word The combination of a code character and numeric value to form an alphanumeric command value to initiate some function or movement on CNC systems.

Word address format Coding format that uses one or more alphanumeric characters to specify the meaning of the programmed word.

X address Code word used to describe motion along the X axis as a coordinate value.

Y address Code word used to describe motion along the Y axis as a coordinate value.

Z address Code word used to describe motion along the Z axis as a coordinate value.

Zero offset MCU characteristic that allows the local zero reference point to be moved to a convenient location while the MCU retains the location of the permanent origin.

Zero reference point The origin or fixed location from which all absolute points are designated.

Zero shift MCU characteristic that allows the zero reference point to be shifted over a specified range. However, the MCU does not retain the location of the permanent origin.

Zero suppression The elimination of nonsignificant zeroes to the left (as in leading zero suppression) or to the right (in trailing zero suppression) of significant digits.

Answers to Odd-Numbered Questions and Problems

Chapter 1
3. CNC equipment is equipped with on-board computer systems.
5. Stepper motors, AC servos, DC servos, and hydraulic servos.
9. a. Perform functions impossible or impractical by other methods.
   b. Increase the accuracy and repeatability of parts.
   c. Reduce production costs.
   d. Increase production levels.

Chapter 2
1. a. Cutting speed—the velocity at which a point on the edge of the tool or cutter travels in relation to the workpiece.
   b. Spindle speed—the number of revolutions made by the tool (milling machines) or workpiece (turning machines) per minute, independent of the tool selected or workpiece dimensions. Spindle speed is measured in revolutions per minute (RPM).
   c. Feed rate—the velocity or rate at which the tool or cutter moves into the workpiece.
3.
$$\text{Spindle speed (RPM)} = \frac{CS \cdot 12}{\text{Diameter} \cdot \pi} = 1019 \text{ RPM}$$

5. Feed rate (IPM) = RPM · CL · #teeth = 8 IPM
7.
$$\text{Spindle speed (RPM)} = \frac{CS \cdot 12}{\text{Diameter} \cdot \pi} = 191 \text{ RPM}$$

Feed rate (IPM) = RPM · CL · #teeth = 0.55 IPM

9. Because tool lengths vary, the MCU must be told which tools and how much each tool varies in length in order to maintain part geometry.
11. The appropriate G code is given, along with the specified offset based on the tool radius or tool diameter. The tool will then ramp on or off of the part when compensation is turned on or off, respectively. The compensated position is based on the next programmed tool movement.
13. TNR compensation allows the maintenance of part features when the cutting tool has a cutting point radius rather than a single cutting point.
15. Adaptive control is a technology that allows the MCU to monitor and adjust machine functions based on specified machining parameters.
17. Provide more efficient and productive machining conditions through the constant monitoring and optimization of machining parameters.
19. a. Pivot insertion
    b. Two-axis sweep
    c. 180° rotation
    d. Spindle direct
    e. Turret head

Chapter 3

1. An axis is a direction of possible programmed machine movement.
3. In point-to-point operations, the tool is not in constant contact with the workpiece. The tool travels from location to location above the part surface and then performs the proper operation at that location. In linear-cut operations, the tool is in constant contact with the workpiece, but is limited to straight-line motions. In continuous-path or contouring operations, the tool is not limited to straight-line cuts.
5. N—Sequence numbering
   G—G codes
   X,Y,Z—Primary motion dimension words
   I,J,K—Auxiliary motion dimension words
   U,W—Secondary motion dimension words
   P,Q—Dwell times and canned cycles
   D—Depth of cut and dwell
   F—Feed rate
   S—Speed

T—Tool information

M—Miscellaneous codes

7. Zero shifting allows the relocation of the zero reference of the workpiece to another valid location within the travel limits of the machine.

9. N10 G0 G90 X-5 Y0 F18 S2000 T1 M6
   N20 X2 Y4 Z.1
   N30 G1 G91 Z-.6
   N40 Z.6
   N50 X2 Y-1
   N60 Z-.6
   N70 Z.6
   N80 X2 Y-1
   N90 Z-.6
   N100 Z.6
   N110 G0 G90 X-5 Y0 M2

11. /G28 U0 W0
    /G0 U-2 W-1
    /G92 X7 Z5
    N10 G0 G90 X5 Z3 G95 G96 F60 S1200 T101 M13
    N20 X1.75 Z.1
    N30 G1 Z-2
    N40 X2.1
    N50 G0 Z.1
    N60 X1.5
    N70 G1 Z-2
    N80 X2.1
    N90 G0 Z.1
    N100 X1.25
    N110 G1 Z-2
    N120 X2.1
    N130 G0 Z.1
    N140 X1.125
    N150 G1 Z-2
    N160 X2.1
    N170 G0 Z.1
    N180 X0
    N190 G1 Z0 F80
    N200 X1
    N210 Z-2
    N220 X2.1
    N230 G0 X3 Z1 T100 M30

13. /G28 U0 W0
    /G0 U-2 W-1
    /G92 X7 Z5
    N10 G0 G90 G95 G96 X5 Z3 F60 S1200 T101 M13
    N20 X1.75 Z.1
    N30 G1 G91 Z-2.1

```
N40 X.35
N50 G0 Z2.1
N60 X-.6
N70 G1 Z-2.1
N80 X.6
N90 G0 Z2.1
N100 X-.85
N110 G1 Z-2.1
N120 X.85
N130 G0 Z2.1
N140 X-.975
N150 G1 Z-2.1
N160 X.975
N170 G0 Z2.1
N180 G90 X0
N190 G1 Z0 F80
N200 X1
N210 Z-2
N220 X2.1
N230 G0 G90 X3 Z1 T100 M30
```

Chapter 4

1. Common tape materials include paper, oil-resistant paper, Mylar-paper laminations, and aluminum.

3. Availability, cost, reusability, standardization, and ease of storage.

7. Parity is an equality of condition between subsequent values. In other words, the same condition or state exists for all values. In part programming, parity checking determines if each byte contains an even or odd number of bits. It is important to check for parity to help eliminate errors or missed-typed information.

9. a. Fixed sequential requires that all values appear in each block and that each block is the same predetermined length.
   b. The block address format eliminates repeating redundant information, but requires the use of a change code to specify which values change between blocks.
   c. Tab sequential format uses tab codes to separate data values within the block, but has the same limitations as the fixed sequential format.
   d. Word address format is the most popular programming format. Letter addresses identify programmed values, but these addresses must appear in the order specified in the format detail for the MCU.
   e. The interchangeable or compatible format is the most versatile format. It also uses letter addresses for programmed values, but words can appear in any order.

Chapter 5
1.

$$\sin \theta = \frac{\text{Opposite}}{\text{Hypotenuse}}$$

$$\cos \theta = \frac{\text{Adjacent}}{\text{Hypotenuse}}$$

$$\tan \theta = \frac{\text{Opposite}}{\text{Adjacent}}$$

3. $\cos^{-1}(.6) = \theta = 53.13°$
   $\sin^{-1}(.6) = \theta = 36.87°$
   90° right angle
5. $\Delta X = \tan(45/2)° \cdot 0.042 \text{(Cutter Radius)} = 0.022$
   $\Delta Z = \tan((90-45)/2)° \cdot 0.042 = 0.022$
7. $\Delta X = \tan 30° \cdot 0.375 = 0.217$
   $\Delta Y = \tan 15° \cdot 0.375 = 0.100$
9. G02/G03 X.... Y.... (Z....) I.... J.... (K).... in word address format.
13.

$$\Delta J = 1.375'' - 0.5 = 0.875$$

$$\Delta I = \sqrt{1.25^2 - 0.875^2} = 0.893$$

$$\Delta X = \Delta I - \sqrt{(R - CR)^2 - (\Delta J - CR)^2} = 0.243$$

$$\Delta Y = \Delta J - \sqrt{(R - CR)^2 - (\Delta I - CR)^2} = 0.236$$

Chapter 6
1. G1 X... Y... Z...
3. I, J, and K are auxiliary words used to describe the centerpoint of an arc in the centerpoint method of circular interpolation.
5. G3 X0 Y3 I0 J3
7. Helical interpolation is used to cut helical pockets and threads in three axes simultaneously.
9. To program a helical thread, three things must be known first: (1) the direction of the thread (right-hand or left-hand); (2) the number of turns of the thread (specified in the part drawing); and (3) the feed rate required to cut the thread.
11. G17 G14 X1.9351 Y0 Z-1 I0 J0 F12 L10

Chapter 7
1. The X and Z axes.
3. G33 Zn_1 Kn_2
5. G71 Pn_1 Qn_2 Un_3 Wn_4 Dn_5 F S
7. G76 Xn_1 Zn_2 In_3 Kn_4 Dn_5 Fn_6 An_7
9. G33 Z1 K.05

Chapter 8
1. a. Locating and resetting the part origin.
   b. Establishing the tool-change/part-change location (unless specified by the tool builder or tool-change mechanism) and moving there.
   c. Measuring and recording the tool number, tool length offsets, and tool diameters used in the program. These values are then entered into the controller's memory.
   d. Appropriate speeds and feed rates for each tool should be recorded. The operator then has the option of using the override features of the machine's controller.
   e. Load the part program into the controller's memory.
   f. Dry-run the part program to check for error-free execution.
   g. Load the workpiece in the machine, making sure that it is firmly clamped in position and that the part origin is located properly, as diagrammed on the setup sheet.
   h. Run the part program.
   i. Fully inspect the first part produced by the program to determine the accuracy of the part.
   j. If the part passes full inspection, production commences. If the part fails inspection, it is necessary to determine how and why the part failed and relay this information back to the programmer.
3. Milling machines and machining centers may be used to drill, tap, ream, bore, counterbore/countersink, and mill or a combination of two or more of these operations.
5. G81 X3 Y5 Z1.1 Z.1 F12
7. G83 X4.5 Y7.25 Z3.1 Z.85 Z.5 F12
9. G84 X1.75 Y4 Z.85 Z.1 F8
11. G86 X5 Y7 Z2.6 Z.1 F7
13. G87 X1 Y1 Z2.85 Z.25 Z.25 F15

Chapter 9
1. CNC punches do not have a rotating spindle or chuck.
3. Punching, notching, nibbling, knockouts, and louvers.

7. The EDM process may be used for any material that is a good electrical conductor. This includes metals, their alloys, and most carbides. Properties such as melting point, hardness, toughness, and brittleness of the stock do not limit the use of the EDM process. EDM can be used to form holes or shapes in materials that may be too hard or brittle to be machined economically by other methods. Since there is no direct physical contact between the tool and the workpiece, very delicate work may be done and very thin materials may be machined.
9. Flame cutters can perform the following operations: contouring, linear cuts, squaring, edge preparation, beveling, and curve fitting.

Chapter 10

1. A loop allows the programmer to jump back to an earlier part of the program and execute the designated programmed movements a specified number of times. This reduces the program length and redundancy of writing these instructions more than once.
3. A subroutine is a small program within a larger main program. Subroutines reduce programming time and the number of required statements. They also allow the programmer to assign values to variable parameters defined within the subroutine.
5. Nesting allows loops to be defined within loops or subroutines and subroutines to be defined within other subroutines.
7. ```
   #1
   N100 G0 G91 X1 Y0
   N110 G1 Z-1.1
   N120 G0 G90 Z.1
   $
   N10 G0 G90 X-5 Y0 F18 S1800 T1 M6
   N20 X0 Y1 Z.1
   =#1
   =#1
   =#1
   =#1
   =#1
   N30 G0 G90 X-5 Y0 M2
   ```
9. ```
   #1
   N100 G0 G91 X.625 Y.625
   N110 G1 Z-.6 F18
   N120 X3.25 Y0
   N130 X0 Y3.25
   N140 X-3.25 Y0
   N150 X0 Y-3.25
   N160 Z.6
   N170 X-.625 Y-.625
   $
   N10 G0 G90 X-5 Y0 F18 S2000 T1 M6
   N20 Z.1
   ```

```
G30 =#1
G31 =#1
G32 =#1
G30 G32 =#1
N30 G0 G90 X-5 Y0 M2
```

Chapter 11

1. Speed, reliability, storage, and standardization.
3. APT and COMPACT II.
5. The postprocessor is a separate computer program that reads the statements input by the programmer, translates these statements into appropriate code, and writes the translated code to an output file.
7. The first phase reads the input file and scans it for errors. Statements are classified and organized according to type of operation. They are then translated into a form ready for the next section.
   Phase 2 is the arithmetic phase. It receives the information from the first section and, using a library of subroutines, tables, and symbols, generates the equations necessary for describing the cutter path for the part programmed. These equations describe the path generated by the center point of the tool in three-dimensional space.
   Phase 3 is the edit phase. In phase three, multiaxis programming, translated cuts, and multiple copy cuts are translated based on output from the second phase.
   Phase 4 is the postprocessor phase.
9. The CAD/CAM link allows programmers to design and submit the part geometry to a postprocessor within the computer environment, where appropriate code is produced. This significantly reduces the development time for parts and allows greater ease in design changes.
11. A flexible manufacturing system is based on hardware and software components that act in a coordinated fashion to produce a wide variety of similar parts. A flexible manufacturing system consists of CNC machines, robots, inspection, and material-handling equipment that can take a part from raw material, perform all necessary machining, material handling, and inspection to produce a finished part.

INDEX

Absolute positioning, 71
Accuracy, 12
AC servos, 13
Adaptive control, 48
Adaptive control with constraints (ACC), 49
Adaptive control with optimization (ACO), 49
American National Standards Institute (ANSI), 87
American Standard Code for Information Interchange (ASCII), 87
APT (Automatic Programmed Tools), 217
   geometry statements in, 225
   postprocessors in, 220
   surface modifiers in, 220
   tool motion statements in, 220
Automatic tool changers, 38
Axis, 57

Binary coded decimal, 86
Binary system, 86

CAD, 16, 224
Call statement, 212
CAM, 16, 228
Cartesian coordinate system, 57
Cast alloys, 25
Cathode ray tube (CRT), 4
Cemented Carbide Producers Association (CCPA), 30
Cemented carbides, 25

Ceramics, 26
Cermets, 26
Circular interpolation, 114, 133
   word address format for, 134
Closed-loop systems, 13
Coated carbides, 26
COMPACT II, 217
   geometry definitions in, 220
   machine tool link, 220
   tool motion described in, 220
Computer-aided part programming, 16
Computer-integrated manufacturing (CIM), 228
Computer Numerical Control (CNC), 12
   advantages, 14
   defined, 3
   disadvantages, 15
   objectives, 14
Continuous path tool control, 62
Coordinate systems, 57
   Cartesian coordinate system, 57
   machining centers, 56
   turning machines, 56
Cubic boron nitride, 27
Cutter offset
   calculations of, 116–125
   defined, 114
Cutting speed, 20, 239

DC servos, 13
Diamond, 27
Direct numerical control, 8

Distributive numerical control, 8, 228

Electromechanical reader, 98
Electronics Industries Association (EIA), 85

Feed rate, 22
  for selected operations, 239
Fixed sequential format, 93
Flexible manufacturing systems (FMS), 15, 228
Floppy disks, 7
Format detail, 66

G codes (RS-274D), 233

Helical interpolation, 137
  word address format for, 139
High carbon steel tooling, 25
High speed steel (HSS) tooling, 25
Hydraulic servos, 13

Incremental or relative positioning, 73
Indexable inserts, 28
Interchangeable or compatible format, 96
International Standards Organization (ISO), 30
Interpolation
  circular
    defined, 114, 133
    word address format for, 134
  helical
    defined, 137
    word address format for, 139
  linear
    defined, 129
    word address format for, 130

Lathes and turning centers, CNC
  axis designations, 56
  common canned cycles, 152–159
  setting the coordinate system, 144
  tool changers, 38
  word addresses, 233

Law of Cosines, 109
Law of Sines, 106
Linear-cut control, 61
Linear interpolation, 114
  defined, 129
  word address format for, 130
Loop systems, 13

Machine control unit (MCU), 2–3
Machine tool link, 220
Machining centers and milling machines, CNC
  axis designations, 56
  common canned cycles, 168–179
  setting the coordinate system, 164
  tool changers, 38
  word addresses, 233
Magnetic media, 7, 84
Manual data input (MDI), 7, 82
Manual tool changes, 37
Mathematics for CNC programming, 237
M codes (RS-274D), 233
Mean Time Between Failure (MTBF), 12

Nested loops and macros, 202
Nesting, 189
Numerical control, 11
  advantages, 14
  defined, 2
  disadvantages, 14
  objectives, 15
Numerically controlled lathes
  axis designations, 56
  common canned cycles, 152–159
  setting the coordinate system, 144
  tool changers, 38
  word addresses, 66
Numerically controlled milling machines
  axis designations, 56
  common canned cycles, 168–179
  setting the coordinate system, 164
  tool changers, 38
  word addresses, 67

INDEX

Oblique triangles
　defined, 109
　solutions for, 110–112
Open-loop systems, 13

Parity, 87
Parsons, John, 10
Part program, 6, 63
Photoelectric reader, 99
Point-to-point programming, 60
Polar coordinates, 112
Positioning systems, 71
Positioning tolerance, 12
Postprocessor, 217
Precision surface sensing equipment, 165
Programming
　absolute positioning, 71
　APT
　　geometry statements in, 225
　　postprocessors in, 220
　　surface modifiers in, 220
　　tool motion statements in, 220
　CAD/CAM, 224
　Circular interpolation
　　defined, 114, 133
　　word address format, 134
　COMPACT II
　　geometry definitions in, 220
　　machine tool link, 220
　　tool motion described in, 220
　examples, 244
　helical interpolation
　　defined, 137
　　word address format for, 139
　incremental positioning, 73
　linear interpolation
　　defined, 129
　　word address format for, 130
Program transfer through RS-232 port, 7–8, 100–101

RAM (random access memory), 3
Rectangular coordinates, 57
Reliability, 12
Repeatability, 12

Right triangles
　defined, 105
　solutions for, 106–107
ROM (read only memory), 3
RS-232, 100–101
RS-244, 87
RS-358, 87

Servomechanisms
　AC servos, 13
　DC servos, 13
　hydraulic servos, 13
　stepper motors, 13
Setup sheet, 63, 64
Spindle speed, 20
Stepper motors, 13

Tab sequential format, 92
Tape formats, 85
　RS-244, 87
　RS-358, 87
Tape punch/reader units, 97
Tape specifications, 83
Tool changers, automatic, 38
Tool length, 43
Tool length offset (TLO), 43
Tool nose radius, 47
Tool radius or tool diameter compensation, 46
　word addresses and codes for, 47
Tool rake, 31
Tooling for CNC operations
　carbide, 25
　cast alloys, 25
　cemented carbides, 25
　ceramics, 26
　cermets, 26
　coated carbides, 26
　cubic boron nitride, 27
　diamond, 27
　high carbon steel, 25
　high speed steel (HSS), 25
　inserts
　　identification, 32–37
　　shapes, 31, 34
Trigonometry for CNC programming, 237

Turning machines, CNC
   axis designations, 56
   common canned cycles, 152–159
   setting the coordinate system, 144
   tool changers, 38
   word addresses, 233

Word address format, 66, 92
   circular interpolation and, 134
   helical interpolation and, 139
   linear interpolation and, 130
   loops and, 207
   mirror imaging and, 208
   subroutines and, 207
   tool radius compensation and, 47
Work cell, 230

Zero shift, 75
Zero suppression, 91